An Introduction to Practical Animal Breeding

Second edition

D. C. Dalton

COLLINS
8 Grafton Street, London W1

Collins Professional and Technical Books
William Collins Sons & Co. Ltd
8 Grafton Street, London W1X 3LA

First published in Great Britain 1980 by Granada Publishing Limited
Reprinted 1981
Second Edition published by Collins Professional and Technical Books 1985

Distributed in the United States of America by Sheridan House, Inc.

British Library Cataloguing in Publication Data
Dalton, D. C.
   An introduction to practical animal
   breeding.—2nd ed.
   1. Livestock—Breeding
   I. Title
   636.08′21   SF105

ISBN 0-00-383025-X

Typeset by V & M Graphics Ltd, Aylesbury, Bucks
Printed and bound in Great Britain by
Mackays of Chatham, Kent

Contents

Foreword

by Professor John W. B. King, Head of the Agricultural and Food Research Council's Animal Breeding Liaison Group, School of Agriculture, Edinburgh

The need for better understanding of animal breeding has probably never been greater. All countries with agricultural interests aspire to recording schemes for the improvement of their livestock, yet rarely have these schemes kept pace with a general understanding of what animal improvement is about, its scope and limitations. From the early days of genetics there has been no lack of books attempting to show how the science of genetics could be put to practical use but few have stood the test of time. This book therefore fills a long-standing gap in the application of genetics to animal breeding and should make clear to students and practical breeders alike the general principles of the subject and the methods by which practical changes in animal performance can be created.

The science of genetics now has a long history but its practical application still lags for a variety of reasons. The logical steps between Mendel's observations on his peas and improving the profitability of a breeder's animals can be tenuous and it is hardly surprising that many authors and students have lost their way on this long road. The principal difficulties of the learner appear to be the technical jargon of the subject, the gap between Mendelian and population genetics and the appreciation of what characteristics contribute to greater profit in farm animals. Dr Dalton has successfully covered these fields and should take students and breeders along with him.

It is refreshing to see a description and discussion of the various traits in farm animals in which improvement is sought. Beyond this the great strength of this book is in the bridge it establishes between Mendelism and population genetics and the way in which the key subject of variation is treated. The number of students who have turned away from the subject because of its mathematical and statistical content must be legion. As this book will hopefully show, these fears are unnecessary and mathematical skills are not essential to a better understanding of the subject.

The more practical minded will in particular welcome the section on breeding and practice. Some may be disappointed to find no general solutions to their problems but, on re-reading, may come to appreciate that better understanding of the principles will provide them with their own answers. This is a thoughtful and useful book which will undoubtedly find many applications and is written in a way which should appeal to a wide audience. The reader should be stimulated to discover from this new exposition that some of the difficulties of the subject he imagined were unreal and be delighted to find that the simple ideas which emerge can be put to very good practical use.

J. W. B. King

Preface

The object of this book is to explain the principles of genetics and animal breeding to students who will eventually be involved in the diverse practical aspects of animal production. To achieve improvement, the modern breeder needs to have a knowledge of genetics, which is the science of heredity. However, genetics can be difficult to understand because it is not a descriptive science like botany or zoology – it is more like chemistry and mathematics, where each step has to be understood before proceeding to the next.

Recent developments in genetics have made the subject more complex than was seen in Mendel's initial work (first reported in 1865). It seems to be the gap between Mendelian genetics and the quantitative genetics of farm animals that students find difficult to negotiate. There is now a wealth of scientific literature on various aspects of the subject, but books and journals containing the relevant information are often difficult for students to locate and are often not easy to understand because they assume a basic knowledge of the subject.

This book explains in simple terms the essential principles and directs those students who need further information to the appropriate texts. The book is presented in five parts: part one deals with the traits or characters of farm animals that concern breeders, part two covers the fundamentals of genetics (Mendelism), while part three carries this through to population genetics and selection. Part four discusses the methods by which practical improvement is carried out, while part five provides further practical advice on how to collect records and then how to plan and publicise the end product of a breeding programme. The book gives references to the major works on animal breeding but references are not given to the many scientific papers on each aspect. Details of these can be obtained from the major textbooks. It is hoped that this book will provide a practical background to animal breeding and that it will stimulate a greater interest in the subject.

Acknowledgements

Thanks are due to many people for help in producing this book; to the publishers for recognising the need for such a text and to their advisers for valuable help with the manuscript. These were Dr M. B. Willis, Mr R. G. Johnston and Dr J. W. B. King who also kindly wrote the foreword. Permission to copy diagrams was kindly given by Professor D. S. Falconer.

Thanks are also due to many colleagues in New Zealand for comments and use of material. I wish especially to thank Professor A. L. Rae for discussion and guidance. Among others were Professor A. R. Sykes, Drs R. L. Baker, J. N. Clarke, J. L. Adam and M. L. Bigham, Mrs Clare Callow and Messrs D. G. Elvidge, V. R. Clark and G. L. B. Cumberland. Typing the drafts was done by my wife, assistance with diagrams by Miss Pauline Hunt and photographic work by D. H. B. McQueen. Specimens of bull semen were collected and prepared by staff of the New Zealand Dairy Board.

D. C. Dalton
Hamilton, New Zealand.

I The Traits in Farm Animals

Man and his animals

Most of the animals currently husbanded by man were domesticated in neolithic times with the exception of the dog which was used in the earlier paleolithic era. Few further attempts have been made in recent times to domesticate animals except for the Eland. Most effort seems to have gone into improving the animals already in use. The modern farmer can improve animal performance in many ways. He can feed his stock better, improve their physical environment by housing, reduce the ravages of pests and disease and so on. These are management or environmental improvements and should go hand-in-hand with better breeding or genetic improvements.

Man's association with animals has always been complex and it remains so. It is wrong to assume that all farmers keep livestock for the same reason i.e. financial gain, and that all breeders have similar aims. This highlights one of the biggest problems in breeding, that of defining the objectives in breeding programmes.

Traits: a general comment

A major difficulty in farm animal breeding is that often breeders try to breed for too many things at once, and are usually disappointed at the slow rate of overall success. It must be accepted that one of the basic principles of breeding is that the larger the number of traits included in a breeding programme, the slower is the rate of progress in any one of them. The main challenge is to decide on a priority order for the required characters, to keep the list short and to stick to this decision. This is where the greatest arguments usually arise – especially between breeders and geneticists. As Lerner and Donald[1] pointed out, any controversy between breeders and geneticists is mostly about aims, less about methods and not at all about theory.

Traits in farm animals can be classified in a number of ways. They can be divided into either simple traits like coat colour or complex traits like growth and survival, or they can be classified as either objective or subjective. Objective traits can be measured in positive terms such as weight, length, area, percentage, etc., whereas subjective traits are measured by scores, grades, proportions, etc., where a person's opinion greatly affects the assessment. Both objective and subjective traits are used in farm animal improvement.

REPRODUCTION

Reproduction is basic to all livestock production but must be very clearly defined as a trait to be considered by breeders. In the female, the breeder is concerned with a considerable number of different ways of measuring reproductive merit. Here are some examples:

(a) The number of eggs shed from the ovary (ovulation rate).
(b) The number of fertilised ova implanted in the uterus.
(c) The number of dams pregnant per 100 joined with the male, or per 100 inseminated. This may be called the pregnancy rate. Pregnancy may be diagnosed at a standard number of days (e.g. 60) after mating or insemination.
(d) The number of offspring born per animal giving birth or per animal joined with the male. Here some breeders may measure total (live + dead) offspring born, whereas others may use only live offspring born. Live offspring born per birth is often called 'litter size' in pigs and sheep.
(e) The number of offspring castrated (testicles removed) or docked (tail removed) at standard ages.
(f) The number of offspring weaned from the dam at standard ages such as six months for beef calves, four months for lambs and three, six, or eight weeks for pigs.

The terms fertility and fecundity are often confused and their definition may vary throughout the world. Generally the term fertility is restricted to points (a) to (d) above and fecundity to points (e) and (f). Fecundity generally includes aspects of rearing ability. However, the point to stress here is that clear definition is needed for whatever measure of reproductive performance is used.

In the male, fertility covers aspects of quantity and quality of the sperm (spermatozoa) produced. Here motility is important as sperm have to move through the female reproductive tract to fertilise the ovum. The proportion of live to dead sperm or the proportion of abnormal to normal sperm is also noted as other factors affecting fertilising ability of the male (see fig. 1). Characteristics of sperm are especially important in artificial insemination (A.I.) where sperm are collected, concentrated, deep frozen, thawed, diluted and then used in low concentrations. The final merit of sperm is measured by the pregnancy rate of the females inseminated.

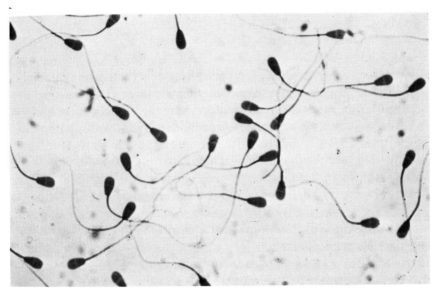

Fig. 1 (a)   **Normal bull semen (× 500)**

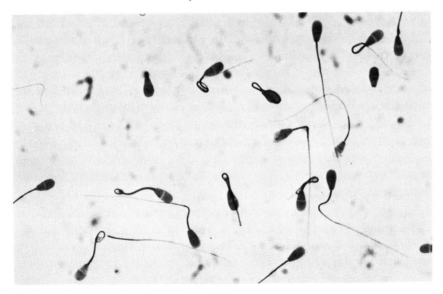

Fig. 1 (b)   **Defective bull semen (× 500)**

In natural service, perhaps more so than in artificial insemination, the libido or sex drive of the male can be critical to the final pregnancy rate. Libido is of special concern to breeders in difficult environments such as the very hot and humid conditions of the tropics or the severe cold of the great plains. The desire of the male to seek and serve females in oestrus (heat) should always be a primary instinct.

BIRTH TRAITS AND SURVIVAL

Breeders are especially concerned with the animal's ability to survive: the more animals that survive then the more there are to provide potential for improvement. Birth and the first three or four days of life are the most hazardous times. First there are the mechanical problems of the birth process in which the offspring has to pass through its dam's pelvis, break free from the birth sac and amniotic fluids and then breathe without suffocating. It has also to withstand a large temperature drop from the dam's body temperature to perhaps ice and snow on a winter range. The most common causes of death in farm animals in the early stages of life are dystocia (difficult delivery) and starvation/exposure. The number of offspring born to the dam at any one time also affects survival: for example, single-born offspring have better chances than individuals in litters.

The breeder often finds it helpful to apportion blame for mortality, which can be classed very simply as: dam's fault, offspring's fault and unspecified (i.e. not sure). Even this presents some problems in assessment and depends a lot on the stockman's opinion. However, the information is valuable because although faults in the offspring can be due to the sire, the faults caused by poor mothering cannot be blamed on the sire; and the true position becomes clear when the sire is mated to different dams. In assessing mothering ability, faults of the dam would be under scrutiny while factors classified as the offspring's fault would be ignored. The complexity of these traits is well recognised by breeders but great efforts are justified in improving them. If a calf dies at birth, not only will the nine months' care during pregnancy have been wasted but the whole twelve months' investment in the dam will have brought no financial return.

Diagnosing the cause of death by post-mortem examination requires considerable expertise and usually back-up laboratory servicing. An accurate diagnosis of the cause of death is often difficult, particularly if the dead animal has not been examined promptly after death. Even apparently simple matters like accurately defining an abortion and a premature birth can be hard. The stockman usually assumes that an abortion has occured if the animal is not born fully-formed, but this assumption is prone to error.

Because of these problems, many breeders take the more positive approach of being more concerned about survival than mortality. This means that they are more interested in why the living offspring live, than why the dead ones die. They thus must select *for* survival characters rather than *against* mortality ones.

Birth is also the time when breeders usually establish the correct parentage of an offspring – the dam can be seen, and the sire to which she became pregnant is known from the records. Problems can arise where dams give birth to their offspring together e.g. under range conditions. Here through mixing of birth fluids and hence the smell of all the newly-born offspring, dams may suckle any of the young. True parentage cannot be determined by the stockman and blood typing is the only guaranteed way to establish parentage accurately. Range cattle often leave their calves in groups or crèches while the dams graze, and as the calves do not usually run to their dams like lambs do, establishing parentage after birth can be difficult. This is an example of how animal behaviour can have profound implications for genetics.

MATERNAL ABILITY
Good maternal ability or 'mothering' is essential in farm stock that suckle their own offspring. It is a complex trait closely associated with survival, as a young animal's apparent desire to live is strongly affected by its dam's ability to feed, shelter and perhaps protect it from predators. Although stockmen can readily recognise good and bad maternal ability in an animal, the trait is difficult to describe objectively. Because of this, breeders often use indirect measures of mothering ability such as the total weight of the offspring at weaning.

Milk production is obviously an important part of mothering, and is under hormonal control along with the processes of reproduction and birth. It is well recognised by stockmen that dams which have a poor milk supply when they give birth also have poor mothering instincts and may not be interested in their offspring. This may be a problem with young dams at the birth of their first offspring. While considering these traits, the breeder must remember the importance of the level of feeding of the dam during the later stages of pregnancy and in lactation measured by her live weight and condition: these are environmental factors.

Some take the approach of deliberately not assisting their animals at birth or up to weaning so that they can identify those dams with good natural maternal ability. It is argued by these breeders that generations of 'good husbandry' (by assisting animals at birth and up to weaning) have retained such defects or weaknesses in livestock. Their approach is one of 'easy-care' where the animals look after themselves and is very important under conditions where labour is expensive.

LACTATION

Lactation requires special attention in farm animals whether they suckle their own young as in beef cattle, sheep and pigs, or whether they are used for milking as in the dairy cow, the dairy goat, and in some countries milking sheep. Lactation involves the whole complex reproductive system which is under intricate hormonal control. Mammary (udder) tissue develops during pregnancy and is ready to function to coincide with birth. Without pregnancy there can be no effective lactation.[2,3] The survival of the young animal is highly dependent on whether or not it receives the colostrum from its dam. Colostrum is the first milk from the udder – it has a thick creamy consistency and is especially rich in antibodies (defence mechanisms) built up against disease organisms by the dam during pregnancy. The intestine of the newly-born animal can only absorb these in the first hours of life.

The udder is of great importance to breeders. In animals that suckle their own offspring it is highly desirable that the teats are of a suitable size and shape to allow the young to suckle and can stand up to the considerable chewing and wear that they get, especially where litters are involved. In such animals, the number of teats is important too. Large teats that young animals cannot get into their mouths are especially bad – the udder pressure builds up causing stress and disease, and the offspring may die of starvation. As the udder in the dairy animal has to hold large volumes of milk at peak lactation, its attachment through the suspensory ligaments to the pelvis is important. With repeated lactation, poorly attached udders become pendulous and are easily damaged when the animal walks or is housed in close confinement with others.

Modern machine milking systems require rapid release or 'let down' of the milk by the dairy animal. This 'let down' mechanism is called a conditioned reflex whereby the animals can be trained to let down their milk. Training is usually done by handling or washing the udder, or giving the cow some extra feed while milking. Fear or stress can effectively 'switch off' the let down hormone. Breeders are thus very concerned to conserve and improve these dairy traits. In some countries, milk is obtained from animals by milking while the calf is suckling or with the calf tied up near the cow's head so she can see it.

Milking machines have caused breeders to pay attention to teat shapes which are suitable for efficient milking but also prevent damage by 'over-milking'. Over-milking is said to occur when the machine continues to squeeze the teat when no milk is left. As milking becomes even more automated in future the physical form of the udder and teats will become even more important.

In the milking animal that has to walk to obtain food and to the milking shed twice daily, overall conformation is important. This involves the udder and teats, large pelvis, good legs and feet, large body capacity for food digestion organs, etc. The breeder of dairy animals is concerned with the quality as well as the quantity of the milk produced. Milk is a complex product and breeders are interested in many of the physical traits such as size of fat globules and chemical traits such as fat, protein, sugar and mineral content.

GROWTH AND DEVELOPMENT

Growth and development are given high priority in breeding. 'Growth' is best visualised as an increase in weight and or size. Sometimes size is inferred from the weight but this can be misleading. 'Development' is more the change in proportion of the various parts of the animal seen through changes that start at conception and continue through to maturity.[2,3,4] Growth is the increase in weight and or size that occurs over time (i.e. age) and can be drawn as an S-shaped (sigmoid) curve in fig. 2.

This curve shows that life begins at conception and growth is rapid up to birth and thereafter to puberty or sexual maturity. Puberty is usually taken as the point of inflexion of the curve or where it changes direction. After puberty the rate of growth slows down until final maturity is reached.

The different tissues vary in their priority for the available nutrients.[2,3,4] For example the placenta and foetus have first priority, then the brain and central nervous system followed by bone, muscle and fat. It is the relationship between these last three tissues that breeders aim to alter and control.[4] Breeders are concerned with animals that vary in their mature size and weight (e.g. Angus versus Charolais cattle and Southdown versus Oxford Down sheep), hence the actual rate of growth and tissue composition at any one time can vary greatly.[4] The point to

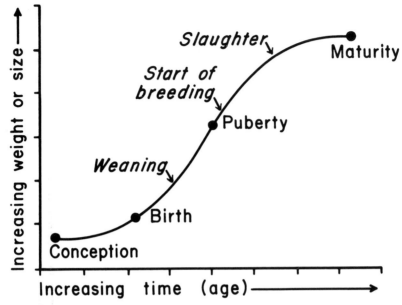

Fig. 2   **Simplified growth curve**

remember in breeding is that all the stages on the growth curve (e.g. birth weight, weaning weight, weight at puberty and maturity) cannot be viewed as isolated traits. If one is altered then the others are affected too.

The live weight of an animal is a simple trait to measure. However, scales do not show what makes up the weight. In ruminants (cattle and sheep) the contents of the digestive tract (gut fill) can account for 10 – 25% of actual weight. So for valid comparisons, animals should be weighed either uniformly full as when straight off feed, or uniformly empty after a standardised period of starvation. Also, in sheep carrying heavy fleeces, variation can be caused by the quantity of water in the wool or whether the sheep were all at the same stage of wool growth when weighed.

THE CARCASS

Breeders are interested in the carcass of most farm animals as it is a stage nearer the consumer than the live animal. However, as the animal has to be killed to examine the carcass, special breeding plans are necessary to select for carcass traits of breeding stock by examining carcasses of offspring or relatives; or ultrasonic aids can be used to study carcass traits on the live animal itself. The weight of the cold carcass as a trait for improvement can be most easily obtained at the point of slaughter. The weight of the carcass as a proportion of the live weight (usually starved live weight) is termed the killing-out percentage (KO%) or dressing percentage.

There are many aspects of the carcass, both objective and subjective, that are important to the consumer and hence to the breeder. The consumer is most interested in the muscle or lean-meat part of the carcass – not the bone or the fat that lies both inside the muscles (intra-muscular) and between the muscles (inter-muscular). Fat can be measured objectively by fat depth at defined points using probes that cut through the external fat layers of the carcass, or sample joints can be minced and a sample analysed chemically. Fat can even be measured by specific gravity – weighing the carcass in air, then in water.

Dissection of the carcass is labour intensive and thus expensive, so it is used mainly in research. Most countries however have grading or classification systems to assess or describe the important commercial aspects of the carcass. These are based on some objective (e.g. fat depth) and many subjective criteria such as distribution of fat cover, fat colour, shape of carcass and proportion of hind end (the first quality or expensive cuts) to fore end (the cheaper cuts). Grading systems are usually criticised for not achieving their aims, but critics often find it hard to suggest workable alternative schemes. Photographic standards are often used in an attempt to retain consistency between graders.

Specially trained taste panels are sometimes used to assess meat qualities after standardised cooking procedures. These include such properties as colour, texture, tenderness, juiciness and flavour. An alternative technique used by some workers is to survey consumers and ask them for their opinions of the product. This technique gives only general information compared to the taste panel, and is similar to that obtained by measuring consumer demand by recording what sells best from the shelves in modern supermarkets.

Modern consumers demand tenderness, flavour, more lean and less fat. It would seem sometimes that breeders are expected to produce an animal that is all hind-quarter! Unfortunately the prospects of altering the proportions of muscles in the body are not high.[4]

The main difficulty facing breeders is to look at a live animal and predict what its carcass will look like or, what is even more difficult, to predict consumers' reactions when they eat the meat. Despite electronic aids that can measure fat depth and eye-muscle area on the living animal, this is still an area for individual skill and experience. Breeders may have to learn to predict live weight (if scales are not available), killing-out percentage and hence carcass weight and grade. Some even attempt to predict the yield of lean edible meat from the carcass. Clearly, there is scope for enormous errors of judgement but these risks will remain until cheap and effective methods are evolved to measure these traits objectively so that the breeder can use them in a programme.

WOOL PRODUCTION

Compared to some of the other traits discussed in Part I, wool appears to pose fewer problems for breeders. However, the greasy fleece as shorn from the sheep is made up of various components that may need the special attention of breeders. The shorn fleece is made up of fibres, water, grease or wax, suint and various contaminants such as marking fluids, vegetable matter and bacterial, fungal and water stains. The breeder has to know which of these can be genetically controlled and which are solely environmental. In most situations, the economic return to the farmer is based mainly on greasy fleece weight.

A fibre is produced from a follicle in the skin of the animal (see fig. 3).

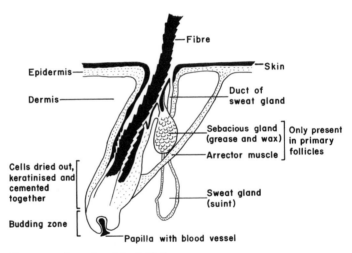

Fig. 3 **Cross-section of a skin follicle**

There are large primary follicles (coded P) that grow coarse fibres and are seen in the British Mountain breeds and some of the hair breeds in other countries. Then there are smaller secondary follicles (coded S) that produce finer fibres as seen in the Merino. Fig. 4 shows the fibres found in two contrasting sheep breeds – Merino and Scottish Black face. The ratio between secondary and primary follicles (called the S:P ratio) dictates the type of fleece produced by the sheep.

There are broadly three kinds of fibres shorn from the sheep and their proportion depends greatly on the breed concerned. There are wool

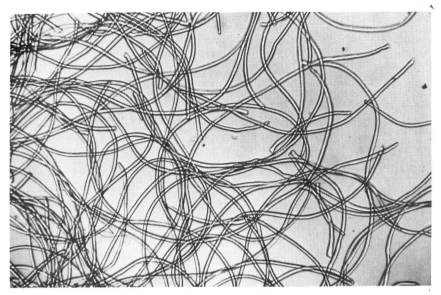

Fig. 4 (a)   **Photomicrograph of Merino wool fibres**

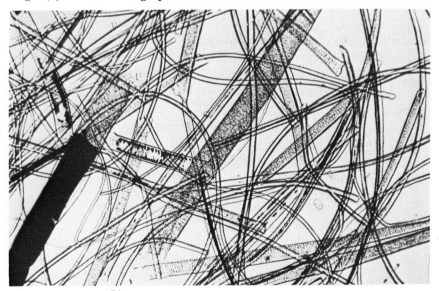

Fig. 4 (b)   **Photomicrograph of Scottish Blackface wool fibres**

fibres that have a solid core, and medullated fibres (or hairy fibres) that have a medulla, or hollow centre, which may or may not be continuous. Kemp fibres may be present and these are notable because they are brittle, they have a medulla and are shed (fall out) from the skin. Fig. 5 is a simplified drawings of these fibres.

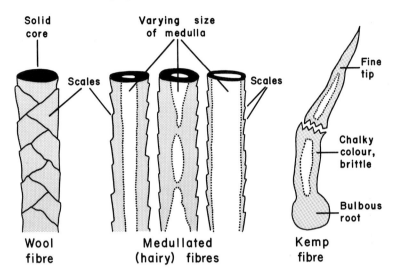

Fig. 5   **Simplified drawing of different types of fibres**

The manufacturer's needs are dictated by the end-use of the wool he buys. The trade's needs are broadly classified into the clothing trade and the carpet trade and each concerns many different aspects of wool such as staple length, fibre diameter, fibre soundness (so that it does not break when put under tension), freedom from contaminants, colour, etc. The traits considered are shown in table 1 on pages 17–19. Technical equipment is continually being developed to measure traits that have traditionally been assessed subjectively; for instance, using fibre diameter measurements instead of the traditional quality number or count. There is now a much greater awareness by breeders of the needs of manufacturers and fashion houses. The International Wool Secretariat (IWS) is actively involved in this area.

DRAUGHT AND SPEED

These traits are of interest to breeders of horses and some cattle breeds in particular countries. Indeed, animal power may increase in importance in future as fossil-fuel supplies decrease and draught power from ruminants that eat fibre still remains a cheap source of energy. Pulling power can be measured objectively by a dynamometer and individual animal performance can be easily recognised. Speed likewise can be assessed by distance travelled in a time period. Despite the fact that speed in racehorses is affected by the official handicapping system and the experience of the jockey, it is still a trait that can be selected successfully by breeders.

Improvement of cattle for draught purposes has had indirect benefits in increased size, improved muscling and reduction of fat. These formerly draught breeds (e.g. Charolais and Limousin) have been widely exploited for beef production throughout the world in recent years.

PHYSICAL FEATURES

Breeders of every class of farm animal are greatly concerned with the physical features of their stock. There is an apparently endless list of traits to consider under the headings of physical form including conformation, structural soundness, visual appraisal, type and many more. Basically they all concern what the animal looks like. They are visual traits, aptly called 'eye-ball traits' by some Americans.

This is a rather difficult area in practice as people usually hold strong views on the importance or otherwise of physical traits. There tends to be a polarisation of breeders at one extreme saying that these characters are important, and geneticists at the other extreme claiming there is no scientific evidence to support that they are related to productive merit. There are many reasons for the contrasting opinions associated with these physical features:

* Many of these traits or their components are difficult to measure in objective terms.
* The way in which some of these traits are inherited is not known.
* Some of the simple traits like colour of hair or wool, presence or absence of horns, may be part of the officially recognised features of the breed association and as such are important to pedigree breeders even if they do not affect the end product – meat, milk or wool, etc.
* As physical traits are assessed by eye, they tend to receive far more attention than do subjective traits seen in the animal's records.

The most difficult part of animal breeding seems to be maintaining a balance between the performance records and visual assessment. However, whether physical traits are based on scientific fact or traditional fancy, they are economic traits. In other words, people pay money for them and they must be considered as such. Why people pay money for characters that are ill-defined but personally satisfying is just part of the wider complex of why people keep animals.

LEARNING ABILITY

Learning ability – sometimes loosely termed intelligence – is most commonly recognised as being an important feature in sheep- and cattle-dogs. Without dogs, many extensive pastoral areas of the world could not be farmed and there seems little chance of them being replaced in the near future. Sheep- and cattle-dogs have learning ability developed to a very high degree and the range of tasks they can execute under command is extensive. The sheep-dog trials held throughout the world vary but the basic tests of gathering, driving and penning sheep are common to all. This means that trial performance can be used to describe the ability of a dog. It is recognised by breeders that luck (chance) can affect a particular trial result, as with a difficult group of sheep that did not herd together. The relationship between the dog and handler is also recognised as important and good dogs are frequently sold on a trial basis until the new owner tests his relationship with a new dog. This applies especially to commands by word (perhaps a different language), whistle or body signal. The dog's learning ability is critical in this.[5]

Breeders are now developing an interest in the ability of farm animals to learn the simple routines that will allow them to do some of their own chores. This is an exciting area for the future and is currently being studied by animal behaviourists and psychologists. Examples of these chores would be operating watering and feeding devices, control lighting, temperature and humidity, treating themselves for external parasites, eating special supplements to counteract deficiencies, operating cleaning out mechanisms, and so on. Economic pressures on breeders will generate a continuing interest in this in future.

TEMPERAMENT

Whatever farm animal is considered and whatever the environment, good temperament is of importance. Breeders have paid great attention to this in the past and modern mechanised systems rely on animal co-operation that comes through a good i.e. non-aggressive temperament. A good temperament in the animal is required in every aspect of the farm routine. Examples are: moving animals around the farm, treating them for ailments, milking, riding, yoking them for pulling and so on. An aggressive animal is a danger to itself (by jumping out of yards), to its fellow animals (by fighting and kicking) and to the people who handle it.

It can be argued that fear of man and dogs can be useful as in mustering animals in extensive situations. If they are not afraid they will not move at the sight or sound of man. It is also argued that fear of man by the animal encourages respect. This is the case with the bull. A hand-reared, over-friendly bull can be a potential danger, but so can a bull that is terrified of his handlers. So clearly somewhere there is a happy medium between the two extremes where the animal is tractable and safe.

The temperament of the animal is controlled by its hormonal condition. At birth an extremely good mother will fight off predators (including the stockman) to protect its young although it is otherwise very friendly. This can be an important quality, as, for example, in sheep protecting lambs against foxes or coyotes.

A quiet temperament is needed in draught animals that may have to spend long periods of time waiting between spells of work. This is often done in noisy, busy conditions and when they cannot eat. In groups of animals there is usually a social hierarchy or 'peck order', in which aggressive animals will push less agressive animals down the order. The higher-order animals usually obtain more to eat, may be milked first, rest first and so on. The lower-order animals eat less, thus produce less and are more prone to disease. These problems are especially important in large groups of animals in confined areas such as poultry in deep litter houses, milking cows in yards or sheep or cattle confined indoors. Stockmen now recognise that there is an optimal group size to reduce these 'stress' factors. However, producing animals to cope with stress situations is a task for breeders and has been especially so for poultry and pigs that are housed intensively.

POULTRY MEAT AND EGG PRODUCTION

Poultry breeders have probably used the recent advances brought about by genetics more than any other of the breeders of farm livestock. Modern poultry are kept solely for their function: the showing of poultry passed into the fanciers' domain many years ago. The very concept of 'breed' in poultry has become outdated in any commercial sense. As the cost of food makes up such a large proportion of the total cost of a bird, the efficiency of feed conversion to meat and eggs is the critical trait in profitability.

Breeders of meat birds (broilers) are interested primarily in growth rate. This can easily be measured and generally the fast-growing bird converts its feed most efficiently. The main management concerns for profit are, then, to eliminate feed wastage by spillage and fouling and to reduce deaths through disease. In meat birds, the breeders are concerned with the conformation of the bird as it appears pre-packed in the supermarket. The bird must have a large breast with fleshy thighs and short legs. Colour of skin can be important for some markets. The bird lays down fat inside its body cavity and this can be easily noticed (and probably removed) by the consumer. Hence attempts are being made to reduce this fat depot so that feed consumed is converted directly to edible meat. Breeders have made great progress towards this end.

In egg production, breeders are concerned with egg number, egg size and weight as well as good hatchability. Body weight and feed conversion of the bird are vital to profit. One extreme would be a small bird that had a low maintenence (feed) cost, laid many medium to small-sized eggs and had a carcass that was of no value at the end of the production period. The other extreme would be a heavier bird that ate more, produced fewer larger eggs but had a good carcass value at the end. Breeders are interested in the most efficient combinations among these traits.

Breeders want birds that mature early so that they start to lay at an early age, that do not pause (stop laying) during production and do not go broody. Persistency is needed too, which is the ability to keep laying for a long time. The various qualities of the egg itself have to be considered as they are important in consumer preference. Such traits are shell colour (white, tinted or brown), shell texture (smooth versus rough), shell thickness as thin shells increase waste through breakages. Inside the egg important qualities are yolk colour, thickness of the white and absence of blood and meat spots.

Summary

The traits discussed so far have been summarised by listing in Table I according to each class of stock. It is important to note that this is by no means an exhaustive list and is not given in any priority order. Deciding what is the priority order is the breeder's main problem and generally it would be based firmly on the economic importance of traits at a particular time. The main concern here is not just the actual value of each trait but their *relative economic value* (REV). This concept is of paramount importance in animal breeding.

Table 1 **Traits considered by breeders of farm livestock**

Class of stock	Objective traits	Subjective traits
Dairy cattle	* Milk yield per lactation of specified length * Fat yield per lactation * Fat percentage * Solids-not-fat percentage * Live wt * Size (withers height) * Lifetime milk yield * Milk flow rate * Calving interval * Heat resistance (heart rate)	* Conformation – udder, teats and structural soundness * Breed specifications * Temperament * Disease resistance
Beef cattle	* Fertility – no. born – age at puberty * Birth wt * Weaning wt * Yearling wt * Wt of calf weaned/cow * Calf weaning wt: cow weaning wt * Mature size and wt * Draught ability * Heat resistance * Fat depth and eye muscle area * Cold carcass wt * Wt of hind: fore quarter * Wt of commercial joints * Bone, muscle, fat wt and proportions	* Conformation – muscling and structural soundness * Breed specifications * Temperament * Disease resistance * Carcass conformation – shape and proportions
Sheep (meat)	* Fertility – no. born * Birth wt * Weaning wt * Yearling wt * Wt of lamb weaned/ewe * Fleece wt and slipe wool production * Milk production	* Conformation – muscling, fatness and structural soundness * Breed specifications

Class of stock	Objective traits	Subjective traits
Sheep (wool)	* Fleece wt * Staple length * Yield * Fibre diameter * Colour * Bulk * Medullation (hairiness) * Follicle parameters – S/P ratio	* Character * Quality number * Break * Cotting
Pigs	* Fertility – litter size – no. litters/ year * No. reared/weaned Litter wt at weaning * Weaning wt/piglet * Wt at slaughter for pork, bacon or heavy wt * Cold carcass wt * Fat depth * Eye-muscle area * Carcass length * Colour of skin * Teat number	* Conformation – structural soundness * Breed specifications * Temperament * Disease resistance
Poultry	* Egg number * Egg size * Egg wt * Hen housed average * Feed conversion – feed consumed per dozen eggs * Egg quality – yolk height * Yolk colour * Shell texture and colour * Body wt * Carcass wt * Proportion of breast meat to total carcass * Shank length * Feed conversion – feed consumed per unit of dressed carcass * Feather colour	* Comb shape * Disease resistance * Aggressiveness

Class of stock	Objective traits	Subjective traits
Dogs	* Working performance – number of trials won * Colour	* Temperament * Eye * Conformation – length of hair for hot or cold environments and feet (pads) * Noise (barking)

II Basic Biology and Mendelism

Basic biology

The cell is the basic component of all living tissue. Its discovery, which led to the 'cell theory', occured in the 1830s and was followed by great advances in biology and genetics. Inside the cell wall is the jelly-like *protoplasm* and in the centre of this is the cell *nucleus*, the control mechanism of life itself. Fig. 6 shows the main parts of the animal cell.

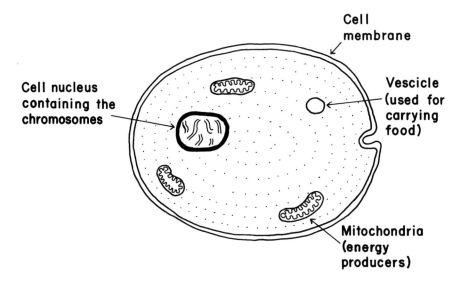

Fig. 6 **The main parts of a simple animal cell**

Within the cell nucleus is the *chromatin* from which the thread-like *chromosomes* develop. On these chromosomes are the *genes* which are the units of inheritence. Genes can be visualised as beads on a string, the string being the chromosome. Fig. 7 is a drawing of what genes or groups of genes look like on the chromosome under a powerful microscope.

Fig. 7   **Drawing of a chromosome showing the dark-banded genes or groups of genes**

In 1953 the chemical structure of a gene was proposed. This is the now well-documented substance called DNA (deoxyribonucleic acid). Biochemical genetics has made great strides in recent years and has progressed into genetic engineering, in which genetic material can be exchanged from one organism to another or perhaps even synthesised in future. There is both excitement and concern over future prospects in this area.

BODY CELLS AND GERM CELLS

Cells are broadly classified into *body* cells and *germ* cells. Body cells are concerned in the main structure of the animal whereas the germ cells are the spermatozoa (sperm) of the male and the ova (eggs) of the female.

Each animal species has a definite number of chromosomes and these are arranged in pairs (called homologous pairs) in the cell nucleus. For example, man has 46 chromosomes (23 pairs); the dog 78 (39 pairs); the pig 38 (19 pairs); cattle 60 (30 pairs); the horse 64 (32 pairs); the donkey 62 (31 pairs); the sheep 54 (27 pairs); the goat 60 (30 pairs). The formation of these chromosomes in the cell nucleus is now a well-documented routine. It is possible to have chromosomes examined (called karotyping) for defects of shape or missing parts. This is especially valuable in human genetic counselling, for example to predict the chances of parents producing mongol children.

When body cells divide in the normal process of animal growth, the chromosomes are halved by splitting down their length into *chromatids*, and equal numbers of these halved chromosomes are drawn to either end of the cell which then constricts between the two new nuclei. This produces two new cells, each with the same number of chromosomes as the parent cells, and the process is called *mitosis*.

In the formation of germ cells or gametes, the process is different from mitosis as the germ cells end up with only half the number of chromosomes that are present in the body cells. This reduction happens during a second division (see fig. 8). This ensures that when a new offspring is formed from the united sperm and egg, it finishes with the correct number of chromosomes for the species. This is called *meiosis*, or 'reduction division'. Fig. 8 contrasts the two processes.

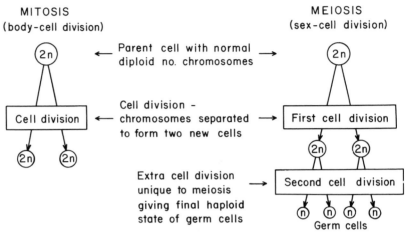

Fig. 8   **Meiosis and mitosis contrasted**

When the female cells divide in meiosis, the first division results in cells of unequal size, the larger divides to produce the ovum and a polar body, while the smaller one produces two more polar bodies. In the division of male cells, nothing is wasted and each half produces a sperm (fig. 9). The challenge comes later for the sperm for although there are millions available to fertilise an ovum, only one actually carries out fertilisation while the remainder die.

The term *diploid* is applied to the double chromosome (or normal) state (e.g. 54 chromosomes in sheep), while *haploid* is used for the germ cells that carry half this number of chromosomes (27 in sheep). A *gamete* is the male or female germ cell. When these gametes combine, the result is called a *zygote*.

CHROMOSOMES AND SEX

There are two broad classes of chromosomes – *autosomes* or ordinary chromosomes, and *sex chromosomes* that specifically control the sex of the offspring. In all species except birds, butterflies and some reptiles, the male determines the sex of the offspring. Thus, in farm animals the sire is the sex determiner. The exception is poultry where this is reversed and the female is the sex determiner.

The letters X and Y are used to describe the sex chromosomes. The dam carries only the X chromosomes (described as XX), and the sire

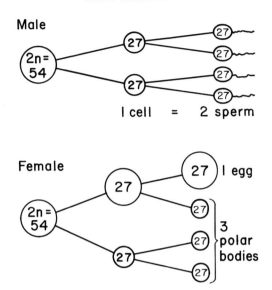

Male
2n=
54
27
27
27
27
27
27
I cell = 2 sperm
Female
2n=
54
27
27
27
I egg
27
27
27
3
polar
bodies

Fig. 9 **Male and female cell division giving sperm and eggs**

carries both X and Y in equal proportions (described as XY). These chromosomes can be recognised by examination with a microscope. Fig. 10 is a drawing from a photograph of the chromosomes of a domestic ram. The photograph has been cut up and the individual chromosomes laid out from largest to smallest. Note the 26 pairs of chromosomes, then the single X and the single Y (much smaller than the X). There are accordingly 26 pairs plus (X + Y), making 27 pairs, or 54 in total.

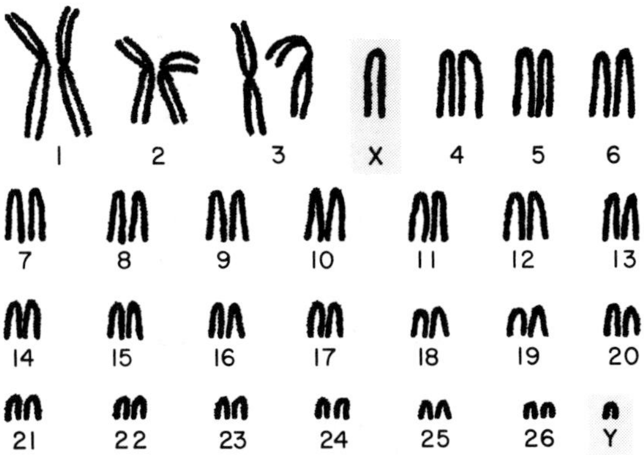

1 2 3 X 4 5 6
7 8 9 10 11 12 13
14 15 16 17 18 19 20
21 22 23 24 25 26 Y

Fig. 10 **Drawing of the chromosomes of a ram showing the 26 pairs and the X and Y chromosomes**

Thus, when a sire mates with a dam the result is this:

| Parents: | Sire | x | Dam |

Chromosomes: XY   XX
                    / \   / \
              segregation   segregation
               ↙   ↘   ↙   ↘

Germ cells: X   Y   X   X
(gametes)   (1)   (2)   (3)   (4)

Offspring: XX   XX   XY   XY
(zygotes)

Sexes: females   males

Note here that the sire is producing X and Y gametes in equal proportions.

The result is males and females in equal proportions but the scientific theory stresses that this is what happens 'on average' and when there are plenty of offspring to test the result. Generally if there is a run of males, this will be counterbalanced by a run of females later. However, there are exceptions to this rule when occasionally sires produce an abnormally high or low number of sons and this may be due to some defective mechanism in the Y-bearing gamete. In general, though, imbalanced sex ratios (i.e. not 50 : 50) are the result of insufficient observations. To work out this ratio, and for the discussion of Mendelism later, it may be advisable to practice how to draw the lines in these crosses to get the right answer and note where the answer is written.

Thus in the example given previously:

(1) is crossed with (3) and the answer written under (1)
(1) is crossed with (4) and the answer written under (2)
(2) is crossed with (3) and the answer written under (3)
(2) is crossed with (4) and the answer written under (4).

Another way of obtaining the answer is to lay it out in a square or checker board where the results of the crosses are put in each box.

Sire

		X	Y
	X	XX	XY
Dam			
	X	XX	XY

CHANGES IN THE NUMBER OF CHROMOSOMES

Normally, the ordinary chromosomes (the autosomes) occur in homologous pairs. However, it is possible to have a situation where more than two chromosomes are present and this is called *polyploidy*. There are a number of different kinds of polyploids, depending on the number of chromosomes present. For example, for three chromosomes denoted by a, b and c, there could be:

Monoploid:	a	b	c
Diploid:	aa	bb	cc
Triploid:	aaa	bbb	ccc
Tetraploid:	aaaa	bbbb	cccc

and so on.

The process of multiplying the number of chromosomes can be done by chemical treatment of the cells and has been exploited in plant breeding where highly productive polyploids are marketed commercially. Viable polyploids are not currently found in farm animals.

CHANGES WITHIN THE CHROMOSOMES

Changes can occur within any one chromosome such as deficiency (where a part is lost), duplication (where a part is added), translocation (where parts of two different chromosomes exchange) and inversion (where parts of the chromosome change). All these changes lead to complications at cell division. (For full information see Sinnot et al.[6] and Strickberger.[7])

Mendel's genetics: Mendelism

The detailed discoveries of Gregor Mendel in his experiments with garden peas in the 1860's have been well documented.[6,7] The main feature for animal breeders to remember is that Mendel's discoveries (termed Mendelism) were based on simple, clearly-defined traits that were inherited as separate entities. These were traits such as colour (either red or white), stem length (short or tall) and skin shape (round or wrinkled) that were controlled by single genes. It is interesting to note that Mendel did record that there were some '... characters that did not stand out clearly'. Perhaps it was fortunate that he avoided working on these and concentrated on single-gene traits. With only a monastery garden as his laboratory Mendel's achievements were prodigious indeed.

One of the major aspects of Mendel's discoveries was to show that the 'choosing' of genes from each parent was controlled entirely by the laws of chance. This profound discovery has been the foundation of all subsequent work and although we know of the exceptions such as linkage, it is still sobering to think of the importance of chance.

TECHNICAL LANGUAGE OF MENDELISM
There is an extensive vocabulary of technical terms that is now part of Mendelism. These terms are fully explained and discussed in relation to current genetics in other texts.[6.7] A few of the terms are discussed here as they are needed to understand the basic concepts of animal breeding.

ALLELE AND LOCUS
The two or more alternative forms of a gene are called *alleles*. The word allele means alternative. The position on the chromosome where they are found is called a *locus* (plural:loci). Unfortunately *gene* and *allele* are sometimes used as interchangeable terms although to do this is not strictly correct. A correct description of the gene for coat colour in cattle for example would be to say that it is present as either the dominant allele black (B) or the recessive allele red (b). Another example is the three haemoglobin blood types in sheep described as A, B and N. This is called an *allelic series* as there can be the following pairs of alleles Hb^A Hb^a; Hb^B Hb^b; Hb^N Hb^n. In cattle blood groups at the B locus more than 250 alleles have been described.

HOMOZYGOTE AND HETEROZYGOTE
A homozygote is an individual that has identical alleles at a specified locus e.g. AA or aa. A heterozygote is an individual that has non-identical alleles at a specified locus e.g. Aa.

DOMINANT AND RECESSIVE ALLELES
A dominant allele covers over or masks the effects of a recessive allele. However, recessive alleles do not always stay covered as they can appear in later generations. There are many examples of dominant and

recessive alleles in farm animals. A few very general examples are
shown in table 2.

Table 2   **Some examples of dominant and recessive alleles**

Animal	Dominant allele	Recessive allele
Cattle	Polled (absence of horns)	Horns (presence)
	Black coat	Red coat
	White head (Hereford)	Other coloured head
	Black & white (Friesian)	Red & white
	Normal size	Dwarf
	Normal palate & pastern	Cleft palate & bent pastern
Sheep	White fleece	Black fleece
	Self colour	Black spotted
	Hairy (medullated) fleece	Non-medullated
	Normal leg length	Short-legged Ancon
	Polled	Horned
	Normally covered	Naked
Horses	Grey coat	Bay coat
	Bay coat	Black coat
	Black coat	Chestnut coat
Pigs	White skin	Black skin
	Lop ears	Prick ears
	Normal nipples	Inturned nipples

The naming of genes

The main system of coding genes is to use letters. Usually capital letters
are given to dominant alleles and small letters to recessive alleles.

Example: P = polled allele: p = horned allele
           W = white allele: w = black allele

In the animal to show the diploid or double state of the alleles, two letters
are used.

Example: PP = polled cow: pp = horned cow
           WW = white sheep: ww = black sheep

Some text books on genetics use mathematical symbols separated by strokes. A plus sign (+) is used instead of the normal allele (sometimes called the wild type) so that if P is regarded as normal, then +/+ is the same as PP. The small letter may still be used for the recessive allele so that +/p is the same as Pp.

In recent years a convention has developed whereby capital letters are used for the gene locus and a small superscript is used to distinguish the allele. Many genes have been named after the person who discovered them or where they were discovered. Examples of the latter are the blood groups discovered in the Lutheran community Lu^a and Lu^b, or the hairy (medullated) gene of the New Zealand Drysdale sheep (N) found on Mr Neilson's farm. This gene N is suggested as an allelic series coded N^j, N^t, N^d and n (the non-hairy). In some cases to avoid confusion, the proposed names for genes have to be submitted to an international indexing body for approval so that the identity of each gene is unique.

Mendel's mathematics and more terminology

To explain the ratios obtained from traits controlled by single genes, the polled and horned alleles are again used. The capital letter P is used for the polled allele. Polledness is the *absence of horns* and is found in cattle, sheep and goats. The polled allele (P) is generally dominant to the allele for *presence of horns* denoted by the small letter p. Note that p denotes presence of horns – it does not control the size and shape of the horns or scurs that may grow. These are controlled by many other genes. However, for simplicity in this discussion p is used to denote horns.

Thus:

P = polled (absence of horns)

p = horned (presence of horns)

PP = homozygous (both alleles identical); dominant (both capital letters present); polled parent.

pp = homozygous (both alleles identical); recessive (both small letters present); horned parent.

In the cross the results are:

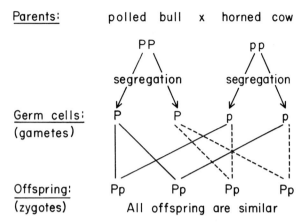

Parents: polled bull   x   horned cow

Germ cells:
(gametes)

Offspring: Pp   Pp   Pp   Pp
(zygotes)   All offspring are similar

Drawn in boxes, it is like this:

	Sire	
	P	P
p	Pp	Pp
p	Pp	Pp

Dam

Pp = heterozygous (alleles non-identical).
Dominant P prevents expression of recessive p so the offspring all
looked polled – *but* they are heterozygous.

Mendelism uses the term *phenotype* to describe what an animal looks
like – its physical form, its colour or its behaviour. The term *genotype* is
used to describe the genetic factors that influence its phenotype. In
farming terms, the phenotype is what you can see in the animal; the
genotype is what is carried in the cells of the animal.

In the previous example, for the

PP animal – phenotype and genotype are the same
pp animal – phenotype and genotype are the same
Pp animal – phenotype looks polled but genotype is not pure polled.

MENDEL'S RATIOS
All the texts on genetics mentioned in the bibliography cover in full what
happens when genes segregate and combine with other genes but the
main points are covered here.

ONE PAIR OF ALLELES

Using the single pair of alleles (polled and horned) on separate chromosomes, the results of segregation are as follows:

SIRE × DAM	OFFSPRING	PHENOTYPE
polled × polled (PP × PP)	PP	All polled – homozygous
horned × horned (pp × pp)	pp	All horned – homozygous
polled × horned (PP × pp)	Pp	All polled – heterozygous

In the crosses between homozygotes (PP×PP) and (pp×pp), the offspring are respectively all PP and all pp. This is often described as 'breeding true to type'. On the other hand, in the mating (PP×pp), the parents do not breed true to type. However, care is needed in using this term 'true to type' for although it may be clear in the Mendelian sense, it is also used by breeders to describe pre-potency or the ability of an animal (usually a sire) to produce offspring like itself and can be more complex. Crossing the two heterozygous polled animals (Pp) gives this result:

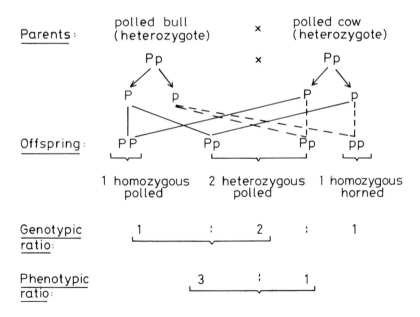

The ratio between the genotypes is 1 : 2 : 1 as shown, but the ratio between the phenotypes is 3 : 1 because the breeder cannot tell the Pp from PP as they are both polled. Thus the Pp parents have not 'bred true': they have bred other types of animals as well as those like themselves.

TWO PAIRS OF ALLELES OF A GENE

With two pairs of alleles on separate chromosomes the situation becomes a little more complicated. An example would be the crossing of Angus and Hereford cattle. The Angus carries the black coat and polled alleles that are dominant over the Hereford's alleles for red body colour and horns. Note that the white head colour of the Hereford is a separate dominant allele. The results of a cross are:

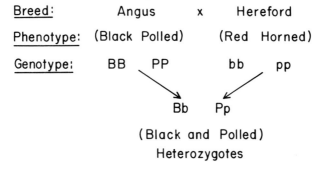

Breed:	Angus	x	Hereford
Phenotype:	(Black Polled)		(Red Horned)
Genotype:	BB PP		bb pp

Bb   Pp

(Black and Polled)
Heterozygotes

When these heterozygous animals are crossed the results are these:

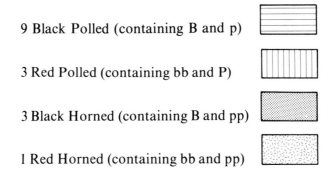

Male

		BP	bP	Bp	bp
	BP	BBPP[1]	BbPP	BBPp	BbPp
	bP	BbPP	bbPP[2]	BbPp	bbPp
Female	Bp	BBPp	BbPp	BBpp[2]	Bbpp
	bp	BbPp	bbPp	Bbpp	bbpp[1]

The phenotype ratio from this is:

9 Black Polled (containing B and p)

3 Red Polled (containing bb and P)

3 Black Horned (containing B and pp)

1 Red Horned (containing bb and pp)

This is Mendel's $9:3:3:1$ ratio. The combinations down the diagonal of the box are important; those at either end marked (1) are identical to the original parents, and the two in the middle marked (2) are new combinations – in this case red polled and black horned. Note that white heads would appear in some of these animals but not in all of them. This would be independent of coat colour and the horned/polled status.

The parent generation of a series of crosses is described as P and the next generation as F_1 or the first filial or daughter generation. When the F_1 generation is bred from, this gives the F_2 or second filial generation and so on to F_3, etc.

MORE THAN TWO GENE PAIRS

Mendel's laws of segregation go further. By using the box layout it can be seen that with four genes the ratios are $27:9:9:9:3:3:3:1$. The situation now is best described by a general formula that says where $n =$ the number of genes, there are 2^n gametes and 3^n genotypes. The value of n is how many times the value 2 or 3 has to be multiplied by itself to give the answer. This is shown as follows:

Number of Genes	Number of Gametes	Number of Genotypes
1	2	3
2	4	9
3	8	27
n	2^n	3^n

Where $n = 20$, the possible number of genotypes is more than a thousand million.

LETHAL GENES

The action of lethal genes causes either the death or physical injury of the animal. Genes that cause injury or maiming may be called semi-lethal as the animal does not always die. Death from lethal genes may occur before birth (in utero or in the shell), in very early life or in later life. Modern surgical skills may now keep animals alive that would formerly have died so the division between lethal and semi-lethal genes may not be definite any more. There are many examples of these lethal genes:[8]

* Dropsical (bulldog) calves in cattle
* Imperforate anus in pigs and sheep
* Hydrocephalus in cattle, sheep and pigs
* Cleft palate (general)
* Nakedness in poultry
* Hairlessness, amputated limbs (general)

Lethal genes that act before birth may be difficult to determine but their presence is usually suspected by the fact that certain predictable genotypes are not seen in the offspring. Care is needed though to ensure that sufficient offspring have been examined before any pronouncements are made, otherwise one could misinterpret chance effects as being the actual cause of the problem. The lethal action of genes can have a 'dominant effect' where it is seen in the phenotype, but the action can also have a 'recessive effect' where the results are not obvious in the phenotype. This latter case is the more difficult to determine. For example:

* For a dominant lethal allele D:
  DD dies; Dd dies; dd lives.
  Here those having the dominant allele (capital D) die.

* For the recessive lethal allele l:
  LL lives; Ll lives: ll dies.

Here, those having the dominant allele (capital L) live and those carrying the recessive allele (small l) die. Ll, the heterozygote, lives because of the dominant allele L, but is a 'carrier' of the recessive lethal allele l.

LINKAGE AND CROSSING-OVER

Some genes that lie on the same chromosome appear to be transmitted as a group. This phenomenon is called *linkage* and the genes are said to be 'linked'. This means that the independent assortment of genes featured by Mendel does not always apply, and an offspring may not get a completely random sample of its parent's genes. The main points about linkage are that while making it harder to get some new combinations of genes it also makes it easier to hold on to existing ones. Lush[9] likened linkage to friction – it can slow up an engine but is very useful in brakes.

Crossing over is the phenomenon of transferring these linked genes from one chromosome of the pair to the other during cell division and is fully described in all texts on genetics (e.g. Strickberger[7]). A simplified explanation of crossing over is as follows.

Consider two parental chromosomes with genes arranged along them. The action happens in the three stages illustrated in fig. 11.

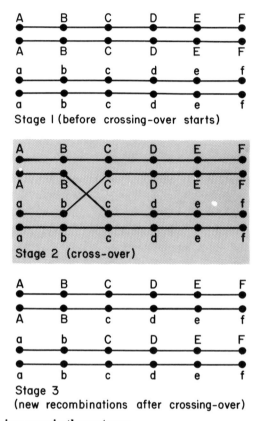

Fig. 11   **Crossing-over in three stages**

The sections of the chromosomes that cross over are called *chromatids*. Crossing over is a breaking-up process among genes that are linked, and tends to bring each pair of alleles into random distribution with every other pair.[9] Thus crossing over has a long-term mixing effect on all the genes.

SEX LINKAGE
Besides determining the sex of the offspring, the X and Y chromosomes can carry genes that affect other traits. This means that the expression of certain characters is affected by the sex of the offspring. This is called *sex linkage* but should not be confused with traits that are *sex limited*. A

good example of a sex-limited character is milk production which males cannot express because of restrictions in their physiology.

Where a gene is carried on the larger X chromosome there may be no corresponding gene on the smaller Y chromosome. This is often drawn by using a bar on the X like this:

$$\text{Male} \qquad \text{Female}$$

$$\text{X Y} \qquad \text{X X}$$

$$\text{or} \quad \overline{\text{X}}\,\text{Y} \qquad \overline{\text{X}}\,\overline{\text{X}}$$

Here the male will show the effect of the single allele of the gene, regardless of whether it is dominant or recessive. It is as if the Y chromosome in the male had no power. In the female, what is actually expressed in the phenotype is what is expected from the genotype. When a gene is carried on the Y chromosome, only the male will show the effect of the gene and it will be seen in the father and the son.

It is very important to know whether the sex-linked gene is dominant or recessive as the situation is different in its transmission. The main practical point is that if the allele is dominant, every affected offspring has an affected parent i.e. it is seen in every generation. If the allele is recessive, the gene appears to 'skip generations'. Two classical examples are severe rickets (a dominant sex-linked gene) and red-green colour blindness (a recessive sex-linked gene) in humans.[10,11]

Sex linkage had been widely used in poultry in the past as an aid to separating the sexes of day-old chicks. Here the sexes could be separated by feather colour rather than by examination of the chick's vent. The colour genes most commonly used in this were the dominant 'silver' allele and the recessive 'gold' allele. Genes for 'barred' colour were also used. These appeared on the wings and back. The term 'auto-sexing' was sometimes used to describe all these genes.[12]

CHROMOSOME MAPPING

It is possible to map the genes on a chromosome using the fact that the closer together two genes are on a chromosome, the less chance there is of an exchange formed between them, and the less chance of recombination occuring. So if heterozygotes are mated and provided there are plenty of offspring (this is no problem in laboratory fruit flies), then the percentage of recombinants is an approximate measure of the distance the genes are apart. By considering a series of crosses involving known linked genes, it is possible to map the chromosomes showing the serial order and approximate distance apart of the genes. Chromosome mapping is usually done with three linked genes. Texts on genetics should be referred to for full details.[7] Although chromosome mapping is well advanced in fruit flies and in maize, it has not yet been exploited in animals but could be useful for the future.

MUTATIONS

A mutation is a change in a gene. Mutations give rise to new alleles at particular loci. Thus mutations are the principal means of producing new heritable variation. There are four main groups of mutations, based on where they happen. These are:

(a) Within the gene i.e. 'intragenic' or 'point' mutations.
(b) Changes in groups of genes on a chromosome.
(c) Changes in the whole chromosome.
(d) Changes in a whole chromosome set.

Studies have shown that genes mutate spontaneously at rates which are generally constant for a particular gene but vary from gene to gene. Mutations can be reversible but the rates of mutation in the two directions are usually very different. A mutation rate for a gene is generally low, for example, from 1 in 100000 to 1 in a million. Albinism in man (a recessive gene) is estimated as occuring 28 mutations per million gametes per generation, whereas deaf-mutism which involves many loci is predicted as 450 per million.[11]*

There is a number of factors that are known to cause mutations such as certain ionising radiations, abnormally high or low temperatures, certain chemicals, other genes carried by the organism or ultra-violet light (non-ionising radiation). Up to the present these effects have only been studied on bacteria, fruit flies and plants, but interest in their effects on animals will increase in future. Increasing radiation and chemicals would be the most likely methods used to cause mutations in animals either through accidental spillages or designed experiments.

There are many examples of mutations in animals. All the examples given earlier as lethal or semi-lethal genes arose from mutations. Some mutants have been made into new breeds or types. The short-legged Ancon sheep is a classical example as are dwarf cattle and horses. Among the wide range in dog breeds are some examples of mutants that have been established as new breeds.

*Note that this is the frequency in gametes, not in zygotes (i.e. children born).

How many genes concern breeders?

Mendel confined himself to single-gene traits and developed his theories for situations involving a limited number of combinations of genes. With farm animals, breeders are generally dealing with thousands of genes and an almost infinite number of possible gene combinations depending on the traits considered. In the fruit fly it is estimated that there are about 6000 pairs of genes on its four chromosomes. Hence in a cow with 60 chromosomes the number of genes must be enormous. Farm animal breeders accept that they are concerned with many genes and this leads into the area of population genetics to be discussed in Part III.

THE ACTION OF GENES: HOW THEY WORK

The way in which genes work is not clear and there are many things still to explain, for example how cells that have similar chromosomes end up as parts of vastly different organs all having different functions. Biochemical geneticists have shown that the main form of gene control is through enzymes. These are the proteins which seem to act as triggers or accelerants to get the function and the animal moving. It also seems that certain genes do not act all the time – they need to be 'switched on' and 'switched off'. Examples would include genes that change coat colours in arctic animals with the approach of winter.

There is evidence that certain genes can act to block certain actions of enzymes and these are called 'genetic blocks'. One example[11] is the dominant allele of a gene that allows some people to taste a group of chemical compounds while some others are 'non-tasters'. Tasting the chemicals is useful clearly as a test for the presence of the gene. The genes involved seem to produce different biochemical side-effects or blocks. Clearly, there must be similar situations in farm animals that have yet to be identified.

PLEIOTROPY

Pleiotropy is a special situation which is found where the same gene has different effects on different traits at the same time. A very good example of known pleiotropy is in the New Zealand Drysdale sheep which has a very strongly medullated (hairy) fleece which is ideal for carpets. However, it has the disadvantage of having very strong horns in the males and smaller horns in the female which cause problems at shearing and in skinning after slaughter. Both the hair and the horns are due to the same N gene so in this pleiotropy there is little chance of selecting for hair while eliminating the horns.

The genetic situation in British hairy and horned sheep breeds has not been so fully investigated as in the Drysdale so it may not be a case of pleiotropy. Indeed, much more is known about pleiotropy in the fruit fly and blood group antigens than about economic characters in farm livestock. A point worthy of note is the contrast between linkage (where gene combinations can break up) and pleiotropy where they do not.

GENE INTERACTION

Gene interaction is said to occur when the same trait is affected by more than one pair of genes (alleles), and these genes may affect each other (interact) in the development of a trait. These different genes are found at many different loci on different chromosomes; and although they may be independent in their segregation, they may not be independent in their action; i.e. they appear to act together to produce the trait. The complexity of the economic traits in farm animals is such that gene interaction must be a major feature although completely documented cases to use as examples are difficult to find.

DIFFERENT TYPES OF DOMINANCE

Genetics textbooks usually discuss a number of different kinds of dominance along with a description of pleiotropy and gene interaction.[13] *Complete dominance* is the first kind and is that already described when the homozygous and heterozygous dominants AA and Aa equally mask the recessive allele aa. *Overdominance* is where the heterozygote Aa performs better than either homozygote AA or aa. This situation is a possible explanation of why some crossbreds or hybrids show superiority in fitness traits. The other kind of dominance is *incomplete dominance* or *no dominance*. These are considered by some authorities to be similar, but others consider them to be different. A good example of this is coat colour in Shorthorn cattle.

The coats of Shorthorn cattle are either red, roan (a mixture of red and white hair) or white. The situation is like this:

$$\text{Dominant red} = RR$$
$$\text{Recessive white} = rr$$
$$\text{Heterozygous roan} = Rr$$

$$RR \times RR = \text{all red (RR)}$$
$$rr \times rr = \text{all white (rr)}$$
$$Rr \times Rr = \begin{cases} 1RR : 2Rr : 1rr \\ 1 \text{ red} : 2 \text{ roan} : 1 \text{ white} \end{cases}$$

Here the phenotypic ratio 1 : 2 : 1 is the same as the genetic ratio – the heterozygote roan can be distinguished from the red. Certain breeds have been founded on these heterozygotes, for example Blue Albion cattle where blue (Bb) is the heterozygote of black (BB) and white (bb). The blue Andalusian fowl is another example.

EPISTASIS

Epistasis acts rather like dominance where one gene masks the expression of the other (both on the same chromosome). Whereas dominance relates to alleles at the same locus, epistasis concerns genes that are *not* alleles. The masking gene is described as *epistatic* to the masked gene. Such genes that do not work in allelic series are very interesting because they modify the size and the direction of each other's effects. Some act as inhibitors and others are described as having 'threshold effects' where they can hold down the expression of other genes.

There are many examples of epistasis, the classical example being feather colour in poultry when two breeds which were once very common in commercial practice are crossed. These are the White Leghorn (WLH) and the White Wyandotte (WW). Both are white but the WLH is genetically a coloured bird with a gene that masks the expression of colour, termed I (a colour inhibitor). This gene is epistatic to C (a colour producer). The WW is a true albino with no colour genes.

The results from the crossing are these:

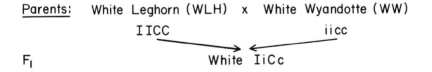

Parents: White Leghorn (WLH) x White Wyandotte (WW)
IICC   iicc
F₁   White IiCc

The offspring has the colour gene but also has the colour inhibitor. Consequently it is white.

Crossing these F_1 birds to give the F_2 shows:

	Male			
	IC	iC	Ic	ic
IC	IICC	IiCC	IICc	IiCc
iC	IiCC	iiCC	IiCc	iiCc
Ic	IICc	IiCc	IIcc	Iicc
ic	IiCc	iiCc	Iicc	iicc

(Female — row labels at left)

9 offspring with I and C = white

3 offspring with ii and C = coloured

3 offspring with I and cc = white

1 offspring with ii and cc = albino

The ratio is 13 white : 3 coloured. Thus epistasis is seen as a phenomenon that alters the classical 9 : 3 : 3 : 1 ratio.

Heredity and the environment

The subject of heredity (genetics) and the environment is a major theme that underlines all work in animal breeding and is discussed in Part III. The main point to make here is that most traits expressed in livestock are due to the influence of *both* heredity and environment working together. The fact that they work together must be stressed. Genes cannot be affected directly by the environment. For some years, it was believed by the Lysenko school of Soviet geneticists that the environment could directly affect genes. The basis of this belief was that if a crop was grown on good land, then the seed from it would have absorbed a 'superiority' (presumed to be genetic) that it would then always have, and it would pass this on to the next generation. Similarly, docking lamb's tails would eventually produce sheep that did not grow tails. Clearly this is not acceptable and it is now universally accepted that the environment cannot affect genes directly. Genes can only provide the messages and direction for the development of the animal's phenotype.The actual expression of the gene is controlled by the environment, for example temperature, light, feeding and management. However, it may be wise to keep an open mind for the future because among the complexities of the environment something may yet be discovered that can directly affect the animal's genotype. This subject is considered fully by Lerner[14].

CURRENT KNOWLEDGE

The 'classical' Mendelian genetics described here, in terms of current knowledge of biochemical and molecular genetics, is an oversimplification.

Between 1940 and 1960, the nature of the gene, its structure and how it worked was studied. In the 1970s amazing developments took place allowing DNA from lower organisms (bacteria and viruses) to be propagated in the laboratory. After 1970 such work became possible in higher organisms including man. This meant that genes could be moved around at will, modified and then put back into living systems. They could in fact be manufactured from chemicals 'off the shelf' and genes could be made that were new to nature.

Some very concerned people now fear this potential and see geneticists as having the power to 'play God'. However, this is certainly an exciting new branch of science referred to as gene surgery where genes can be inserted, deleted or replaced in a living system by genetic engineers so we can now talk of gene machines. Whether these techniques are exploited for good or evil will be seen in the near future.

III Population Genetics, Selection and Breeding

The complexity of traits in farm animals

Most of the economic traits that breeders have to consider in farm animals are not simple and their inheritance cannot be explained by the techniques of Mendelian genetics. Consider for example an apparently simple trait like weaning weight in sheep. The various factors affecting it are shown in fig 12.

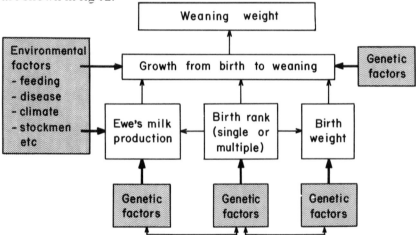

Fig. 12 Factors affecting lamb weaning weight

Here arrows are used to show the links between the various factors. It can be seen that genetic factors are involved in the ewe's fertility, milk production, lamb weight and its growth from birth to weaning – really from conception to weaning. It is also obvious that the environment has an important effect on many of these.

It is in dealing with traits like this that farm animal breeding is concerned with the branch of genetics called 'population genetics'. The terms quantitative genetics or biometrical genetics may also be used. This branch of genetics deals with traits that are controlled by many genes (referred to as polygenic) and where large numbers of animals (or populations) are involved. These populations may be individual flocks or herds, groups within these, or perhaps even the entire national flock or herd. However, although the population becomes the main concept, the individual within the population is still important.

THE BRIDGE BETWEEN MENDELISM AND POPULATION GENETICS

It seems to be at this point of bridging the gap between Mendelism and population genetics that most students have difficulty, and doubtless the explanation of what is going on at the level of the animal's genes has not been clearly explained. Mendelism has not stopped and something else started – the Mendelian actions of the genes carry on. The problem is that explaining their complexity using Mendelian techniques is difficult.

To illustrate this, consider the situation with fleece weight in sheep – taken because it is a fairly simple character compared to many others the breeder has to deal with. *Assume* that it is controlled by two pairs of alleles Aa and Bb. Each of the alleles A and B adds 0.1kg to the basic genotype value of 5.0kg for aabb. Alleles a and b add nothing. So aabb is the base line and it produces 5.0kg of wool. Also, and this is very important, the environment is taken to have no effect. Thus:

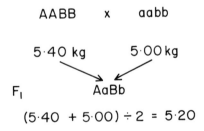

AABB   x   aabb

5·40 kg   5·00 kg

F₁   AaBb

(5·40 + 5·00) ÷ 2 = 5·20

Then when the F_1 animals are crossed the results for the F_2 are these:

No.	Genotype	Fleece weight (kg)	
4	AABB	$5.0 + (4 \times 0.1)$	= 5.4
8	AABb	$5.0 + (3 \times 0.1)$	= 5.3
4	AAbb	$5.0 + (2 \times 0.1)$	= 5.2
8	AaBB	$5.0 + (3 \times 0.1)$	= 5.3
16	AaBb	$5.0 + (2 \times 0.1)$	= 5.2
8	Aabb	$5.0 + 0.1$	= 5.1
4	aaBB	$5.0 + (2 \times 0.1)$	= 5.2
8	aaBb	$5.0 + 0.1$	= 5.1
4	aabb		5.0
64		Average	5.2

This exercise highlights two important points:

(a) The mean of the F_2 is 5.2kg which is the same as the mean for the F_1.
(b) However, in the F_2 a great deal of *variation* has appeared, the fleece weights range from a low of 5.0kg to the high of 5.4kg.

Note also that further variation could be included into these weights by environmental factors and the effect of this would be to increase the spread even more. This then is the link with population genetics: the discussion of variation.

Population genetics: variation

Population genetics starts with the study of variation and asks the question what is causing this variation in a group of animals. Why for example do the weaning weights of a group of lambs vary? In traditional breeding terms, variation in animals appeared to be something to get rid of as quickly as possible and breeders considered success as being able to breed animals like 'peas in a pod'. This was mainly thought of in terms of physical features, i.e. the animals had to look alike and animals that looked different were considered to be inferior. This fact of *looking* alike may not be important in practice but animals that *perform* similarly certainly have advantages: they can be fed, slaughtered and processed in a standardised way and hence have economic advantages.

Geneticists consider variation as the raw material for improvement; variation that is both visible (phenotypic) and invisible (genetic). The breeder's tool to work on this variation is *selection*. So basically, population genetics is all about variation and selection.

VARIATION: HOW TO DESCRIBE IT

Probably the easiest way to understand variation in a flock or herd is to draw a diagram of it (called a distribution) for a particular trait. Good examples would be the live weight at a specific age, fleece weight, milk yield, etc., of a group of animals. There should be a reasonable number of animals in the group, say at least 15 – 20. Fertility and survival should be avoided as these are different and are considered on page 47.

An example used here is the live weight of 2500 young Romney rams at 15 months of age. Firstly, the group sizes were chosen (a 2 kg range was taken), and the number of animals that fell into each class was counted. This is shown in fig. 13 and is called a histogram. It can be seen that the columns are very even on both sides of the middle one which is the average or mean weight of 49.5 kg. A very smooth bell-shaped curve can then be drawn through the tops of the columns to describe the distribution instead of drawing all the columns. This curve, first

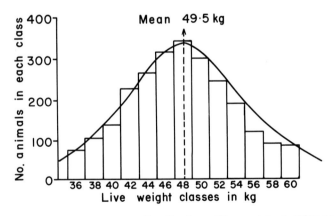

Fig. 13 **Histogram showing the distribution of live weight in 2500 15-month-old Romney rams**

described by the German mathematician Gauss (1775 - 1855) has some well recognised features that are the bases of population genetics (shown in fig. 14). The curve is described as a 'normal' curve or the 'normal curve of distribution'. Its features are these:

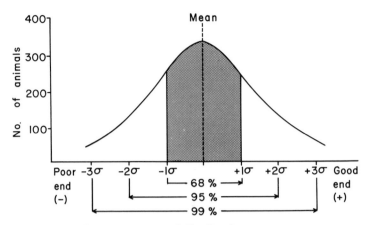

Fig. 14   **The normal curve or normal distribution**

* The curve describes a mean and the variation spread around it.
* The vertical lines drawn from the shoulders of the curve where the slope changes direction enclose an area in which about 68% of the observations will be found. These lines may be called *truncation* points.
* The shaded area can be defined more precisely in statistical terms by saying that it is one standard deviation above, and one standard deviation below the mean (written as mean $\pm 1\sigma$). The Greek letter sigma (σ) is the mathematical notation for standard deviation.
* In the area described by the mean $\pm 2\sigma$ there will be about 95% of all the observations and the mean $\pm 3\sigma$ will cover about 99% of all the observations.
* Thus the mean and standard deviation can quickly describe the variability of a group of animals. This also introduces the concept of describing individual animals by their differences or *deviation* from the mean. This is the above- or below-average concept that is the basis of animal improvement.

OTHER SHAPED CURVES

As stated earlier, the standard deviation is a useful guide to the shape of the curve for a trait and examples are shown in fig. 15. All these distributions described so far have dealt with situations where variation is continuous from one side of the distribution to the other. This situation is very readily recognised on the farm where, for example, there are always some very poor milkers relative to the herd average, the majority are about average and there are some really high-yielding cows. This is called continuous variation.

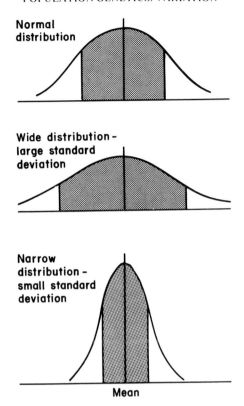

Fig. 15 **Different types of normal distribution**

In some traits, and the best examples are survival and fertility, the variation is not continuous and is referred to as discontinuous or 'discrete' variation. In survival for example, an animal is either dead or alive so there are only two classes. The fertility of cattle or horses, measured by numbers of offspring born, is usually 1 or 0 since multiples are not common. In sheep, pigs or any animals that have litters, however, the distribution could look like fig. 16 where it can be seen that animals either have no lambs (0) or some lambs i.e. anything from one up to six lambs in sheep and many more offspring in pigs. In fig. 16 the mean is about 0.5. This is called a 'skewed' distribution i.e. it appears as a distribution that has been 'pushed over'. These skewed distributions have different statistical features to normal distributions.[13]

THE CHALLENGE : TO IMPROVE THE MEAN
The real breeding challenge in a flock or herd is to improve the mean performance for a trait or a number of traits, and also to try and reduce

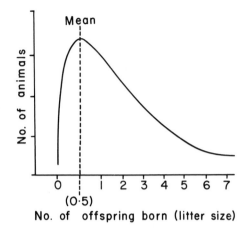

Fig. 16   **A skewed distribution**

some of the variation. Unfortunately in the livestock industry, most of the glory seems to be associated with breeding a few individuals that may win ribbons in the show ring or make a record price. These outward signs of success are easy to feature whereas small increases in mean performance may not be, although on a national scale the latter will give greater economic benefit.

WHERE TO START?

Once an animal's performance has been measured for a character and its position relative to the mean of the whole group established, then the first question should concern the reason for its actual position. This is simply to find out if the animal's performance was due to its genotype or to some environmental factor. Any comparisons among animals assume that they were all treated alike as far as the management would allow.

Consider an example where a decision has to be made on a beef heifer with a poor (below average) 550-day weight. Why is it in the poor end of the herd distribution? The possibilities are these:

* The heifer had a low birth weight because its dam first calved at two years of age instead of three years of age and was also poorly fed before and after calving.
* The heifer's dam gave very little milk and consequently was a poor mother, so the heifer was prematurely weaned at a very light weight.
* Because the heifer was a poor weaner, it was ravaged by worms, insects and ticks and these all helped to restrict growth and hence 550-day weight.

This is a list of 'environmental' reasons and could include many more. However, if this beef heifer were mated even to an average bull, then its progeny could show a definite move upwards towards the mean. On the other hand an animal at the opposite end of the distribution (the +end) could have had all the environmental luck going, and its breeding performance could be much worse than expected. This is a well-known phenomenon in animal breeding and is accepted as part of the hazards of the business.

Statistically this is described as a regression (or return) to the mean. In practice it means that if the superiority of an animal in a population is due mainly to a good environment, then the improvement of mean performance through genetics will be harder. If the expression of the animal's performance is due to good genes, then the breeder can proceed to make decisions with much more confidence that what he sees will be passed on to future generations.

COMPARISONS BETWEEN POPULATIONS

Making comparisons between animals in different populations is difficult and basically should not be done unless it is known where the populations lie on some overall distribution of breeding merit. The reason for this is that because different populations (flocks or herds) are run on different farms with different feeding and management, then these environmental effects reduce the validity of the genetic comparisons between the animals. The comparison thus is between stockmen and not the stock. More technically, even if the deviation of each animal from its flock mean were used in a comparison, that comparison would be invalid because the means are different.

This can be very frustrating for a commercial ram buyer, for example, who may be looking at rams on different farms and may want to compare an above-average ram on a poor farm with an average ram on a good farm. He cannot tell whether he is comparing environments or genetics. This opens up a wide general subject of comparing animals which is discussed on page 135. At this point it is sufficient to say that attempts to get round the problem are either to put all the animals into a 'common environment' for comparison, or to have the same reference animal (used through semen) in each environment. The main point to stress is that comparisons should be made among animals *within* a flock or herd and not *between* flocks or herds.

Selection

Selection can be simply defined as allowing some animals to be parents of the next generation while depriving others of the privilege. Thus castration was one of the earliest forms of selection. What selection actually does is to change the frequency with which certain genes (or combinations) occur in a population. This is the concept of gene frequency that is dealt with fully in all texts on population genetics.[15,16] A hypothetical situation is where all the 'bad' genes have a low frequency or are rare and all the 'good' genes have a high frequency or are plentiful.

However, in practical breeding selection is really about making decisions. It can be defined as 'choice based on information', the information often ranging from solid fact right through to complete guesswork.

ARTIFICIAL AND NATURAL SELECTION

Artificial selection means selection made by the breeder. It contrasts with natural selection where man allows 'nature to take its course'. This natural selection was the core of Darwin's work in which he developed his theory of the origin of different species. In modern agriculture there seems to be very little left that could be described as truly natural so breeders are concerned mainly with artificial selection.

Recently breeders in some countries have made positive attempts to exploit natural selection. These breeders claim that traditional practices have made farm animals 'soft' or have lost 'constitution' – both difficult words to define scientifically. Nevertheless, these breeders argue that animals now seem to require more feed, more drugs and more care than they used to, so they are opting for an 'easy-care' or 'no-care' approach. They are doing this by selecting the animals that survive, making sure that they are unassisted by man in really tough commercial conditions.

This approach has been expanded into selection for disease resistance, for example, by deliberately exposing breeding stock to pathogens (disease organisms). An example is the selection for tick resistance in the Australian Milking Zebu (AMZ) where animals were deliberately infested with a known number of ticks and the ticks that gorged and then dropped off were counted. Poultry breeders have been especially concerned about selecting for disease resistance and have made progress by deliberately challenging the stock with some of the major disease pathogens. However, this whole easy-care approach assumes that the genetic qualities that the breeders want are genetically controlled. At present there is little hard evidence to verify this but plenty of circumstantial evidence from breeders that their approach is effective.

CULLING

Culling is really another word for selection. However, it is used to describe the removal of inferior animals rather than the more positive selection of good ones. Thus selection and culling go together.

It is most important to understand whether the decision to cull has been made for genetic or environmental reasons. It is very easy to cull 'poor looking' stock but genetically this achieves little if they were poor because of environmental reasons. The risks of making this type of error seem to be highest when animals are examined after a period of high production such as a lactation. For example, ewes that have suckled twins are thin and poor-looking while barren ewes are fat. The same would apply to sows that had suckled large litters compared to those that had small litters. It is perhaps understandable that good stockmen like to cull poor looking animals as these tend to reflect on their husbandry skills, but it is important that they appreciate the genetic implications of their actions which may not necessarily be beneficial.

Selection and culling can be visualised as 'pressures' that can be increased or relaxed so that their effects depend on the intensity or strength of the pressure. What selection and culling are doing is altering gene frequency.

Genetic progress – what controls it?

The three factors that control genetic gain in a trait are:

(a) Heritabilty
(b) The selection differential
(c) The generation interval

HERITABILITY

This is the term used to describe the strength of inheritance of a character, i.e. whether it is likely to be passed on to the next generation or not. A precise definition would be : *For a given trait heritability is the amount of the superiority of the parents above their contemporaries which on average is passed on to the offspring.*

Note the key words where care is needed:

* 'superiority of the parents'
* 'above the contemporaries'
* 'on average passed on'

It is *not* how much of the measured trait that will appear in the next generation.

The notation h^2 is given to heritability and is expressed on a scale from 0 to 1.0, or 0 to 100%. It is often necessary to use generalisation such as these:

Low or weak: 0–0.1 (0–10%)
Medium or intermediate: 0.1–0.3 (10–30%)
High or strong: 0.3 or above (30% or above)

Where low becomes medium, and medium becomes high is open to debate – it often depends a lot on the character under discussion and the way one wishes to argue a point. To describe what heritability is, the notation of Professor Lush is used.[9]

The basic equation is this:

$$P = G + E$$

i.e. Phenotypic variation = Genetic variation

+ Environmental variation

$$+ \boxed{\begin{array}{l}\text{Interaction} \\ \text{and association} \\ \text{of G and E}\end{array}}$$

The part in the box is caused by an association that might occur between genetic variation and the environment and the interaction between them.

The (G) and (E) parts can be divided further. First the (G) has three components:

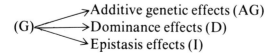

Then (E) has two components:

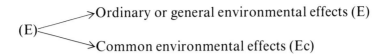

(E) → Ordinary or general environmental effects (E)

(E) → Common environmental effects (Ec)

The additive genetic effects (AG) are the most important part as they are stable and are regularly passed on from one generation to the next. Dominance and epistasis are not passed on with the same guarantee. The environmental part has a special section – the common environmental effect (Ec) which is experienced by members of the same litter as they were all together from conception to weaning and, having some environment in common, are therefore less variable. The main point to note about heritability is that it is a ratio and *not* an absolute value. Because of this, estimates can vary greatly depending on how they are calculated and where the data that were used came from. The number of animals used is also important as the more animals there are the more reliable the estimate becomes.

Table 3 shows some heritability estimates for different traits in farm animals. Note the wide range in values usually found for each trait.

Heritability can be measured in a number of ways:

(i) FROM THE RELATIONSHIP BETWEEN PARENTS AND OFFSPRING

Comparisons of the performance of daughters and their dams can be used in cattle and in sheep. A disadvantage of this method is that the dams are nearly always a selected (and hence biased) group, even if the daughters were not selected. Maternal (mothering) effects may also confuse the situation.

Comparisons can also be made from the relationships between half-sibs . (Sibs or siblings are offspring of the same parents.) Paternal half-sibs are all the offspring by one sire out of different dams i.e. a progeny group, and are most often used to calculate heritability. Maternal half-sibs are progeny from one dam by different sires and are more difficult to obtain except in poultry, pigs or by superovulation of large farm animals. Good estimates of heritability are obtained where there are plenty of sires being compared, each with reasonable numbers of progeny. Minimal numbers would be at least four to five sires with ten progeny per sire, depending on the trait concerned.

(ii) FROM THE ACTUAL RESPONSE TO SELECTION

In some selection experiments where there had been an upward and a downward selection line starting off from a common base, heritability

Table 3 Some examples of heritability estimates in farm livestock (expressed as a percentage)

Dairy cattle	%	Beef cattle	%	Sheep	%	Pigs	%	Poultry	%
Calving interval	0–15	No calves born	0–15	Number lambs born	0–15	Number born/litter	15	Hen housed egg production (day 1st egg to 500 days)	5–10
1st service – conception	7	No calves weaned	0–10	Number lambs weaned	0–10	Litter size at weaning	7	Part records (3–4 months' lay)	25–30
Length and intensity of oestrus	18–21	Calving interval	0–15	Weaning wt	10–40	Mean weaning wt	8	Age at sexual maturity	15–30
Cystic ovaries	15–40	Mature cow wt	50–70	Wt of lamb weaned	30–40	Daily gain	21–40	Egg size	40–50
Milk yield	25–40	Feedlot gain	45–60	10-month (hogget) wt	35	Feed efficiency	20–48	Egg shape	25–50
Fat yield	27–43	Pasture gain	30	10-month oestrus	30–50	Killing out %	26–40	Shell colour	30–90
Udder size and shape	20–40	Efficiency of gain	40	Greasy fleece wt	30–40	Yield of lean %	45	Shell thickness	25–60
Type rating	30–60	Birth wt	20–59	Staple length	30–60	Mean backfat	43–74	Yolk colour	10–40
Peak milk flow	35–86	Weaning wt	20–55	Mean fibre diameter	40–70	Fat depth C	62–65	Albumin firmness	10–70
Average flow rate	67	18-month wt (pasture)	30–55	Crimps per cm	35–50	Fat depth K	42–73	Frequency of blood and meat spots	10–50
Fat %	32–87	Final wt	50–60	Medullation	34–80	Carcass length	40–87	Fertility	0–5
SNF	53–83	Carcass grade	35–45	Greasy colour	20–40	Loin length	39–46	Hatchability	10–15
Protein %	48–88	Rib eye area	70	Birth coat grade	59–80	Leg length	46–50	Viability	1–15
Lactose %	35–62	Tenderness	60	Face-cover score	36–56	Carcass depth	34–56	Body wt	25–65
		Fat thickness	45	Wrinkle score	20–50	Fillet wt	31–54	Shank length	40–55
						Eye muscle area	35–49	Body depth	20–53
						Sex odour in boars	54	Breast width and angle	15–35
								Keel length	30–57

can be calculated from the amount of divergence between the lines. This is not a commonly-used method in farm animals.

(iii) FROM COMPARISONS USING TWINS

Here monozygous twins (one-egg or identical twins) are compared with dizygous (two-egg or non-identical twins). The differences between the performance of identical twins are all environmental. Twins cannot be compared with singles for this study due to the common environment that affects twin pairs. Heritability estimates from twin studies are much higher than from non-twin work. An example is the heritability for milk yield which from field data ranges from 0.20 to 0.39. From twin data the estimates are 0.75 to 0.90.

THE SELECTION DIFFERENTIAL

This a measure of how good the parents chosen to produce the next generation will be. It is the superiority of the selected parents over the mean of the population from which they came. It is an expression of the breeder's aim for a trait. The selection differential can be affected by a number of things on the farm. In a sheep flock for example, if fertility is

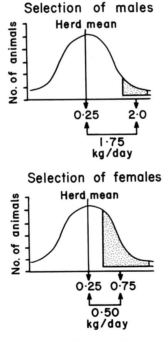

Fig. 17   **Calculation of selection differential**

low and lamb survival poor, and if ewes are drafted (culled) after four lamb crops, then the scope for achieving very superior parents with a high selection differential is very limited. The selection differential can also be reduced if the population is uniform as few animals are far enough above or below the mean to make any impact by selecting the best and culling the worst.

It is important to note that a selection differential can be calculated for both parents – the future sires and the future dams. For some characters like milk yield, the sire's selection differential can only be calculated indirectly from the mean of his offspring. As fewer males are generally needed than females, then a greater selection differential can be applied to them than to females. In some flocks or herds, if performance is low, every available female replacement is required to keep up the numbers so no selection is possible at all as seen, say, in British flocks from the very hard mountain conditions. In this case all the responsibility for genetic gain remains with the sires that were chosen. An example of calculating selection differentials is shown in fig. 17 using gain per day in a herd of beef cattle.

To calculate selection differential for males:

		Gain/day
Mean of selected males		2.00kg/day
Overall herd mean		0.25 kg/day
Selection differential	$(2.00-0.25) =$	1.75 kg/day

To calculate selection differential for females:

		Gain/day
Mean of selected females		0.75 kg/day
Overall herd mean		0.25 kg/day
Selection differential	$(0.75-0.25) =$	0.50 kg/day

These two selection differentials are then averaged to give:

$$\frac{1.75 \text{ (for males)} + 0.50 \text{ (for females)}}{2} = \frac{2.25}{2} = 1.13 \text{ kg/day.}$$

Note what happens when there is no selection of females i.e. where the selection differential equals 0. The calculation then is:

$$\frac{1.75 \text{ (for males)} + 0 \text{ (for females)}}{2} = \frac{1.75}{2} = 0.88 \text{ kg/day.}$$

Clearly the potential genetic gain has been severely reduced from 1.13 kg/day to 0.88 kg/day. The key to success in obtaining a high selection differential is to have plenty of variation to start with, and many more animals than those needed to maintain the flock or herd size so that only the very best can be chosen as parents for the next generation.

It is possible to calculate the actual 'intensity' of selection using this formula:

$$\text{Intensity (i)} = \frac{\text{Selection differential (SD)}}{\text{Phenotypic standard deviation } (\sigma_p)}$$

The phenotypic standard deviation (the standard deviation of the animal's phenotype) is simply a way of describing the variation normally found in the trait for a particular population. Table 4 gives some examples for different livestock.

Table 4 **Some examples of phenotypic standard deviation (σ_p) in farm livestock**

Dairy cattle	Milk yield	254 kg
	Fat yield	43 kg
	Fat percentage	0.48–0.50%
	Days in milk	29–30
Beef cattle	Birth weight	4–7 kg
	200-day weight } Weaning weight	20–26 kg
	Weaning age	18–20 days
	400-day weight	25–30 kg
	550-day weight	25–30 kg
	Pre-weaning gain	0.10–0.15 kg/day
	Feedlot gain	0.10 kg/day
	Pasture gain	0.07–0.10 kg/day
Sheep	Number of lambs reared	0.6 lambs
	Lamb weaning weight	3.6 kg
	Weight of lamb weaned	5.0 kg
	Hogget body weight	4.5 kg
	Ewe fleece weight	0.5 kg
Pigs	Average daily gain	0.06 kg/day
	Feed conversion ratio	0.20 feed/gain
	Fat (C measurement)	2.5 mm
	Fat (K measurement)	2.7 mm

Table 4 *(contd.)*

Pigs	Fat (S measurement)	4.3 mm
	Carcass weight	1.28 kg
	Dressing percentage	1.60%
Poultry	Winter egg production	13.7 eggs
	Spring and summer egg production	23.0 eggs
	Age at sexual maturity	3.7 weeks
	March egg weight per dozen eggs	45 g

The most important aspect to affect the practical breeder is how the number of animals available to choose from governs the progress that can be made. This can be most easily shown in a sheep flock with different levels of fertility shown in table 5 for ewes.

Table 5   **Female selection intensity in sheep**

	80	100	120	140	160	180
Lambing % (lambs weaned/100 ewes joined)[2]	80	100	120	140	160	180
No. females available for replacements (assume 50:50 sex ratio)	40	50	60	70	80	90
No. ewe replacements needed (assume 4-year life in flock)	27	27	27	27	27	27
Percentage selected (No. needed/total available) × 100	68	54	45	39	34	30
Percentage culled (100−% selected)	32	46	55	61	66	70
[1]Expected selection intensity (i)	0.53	0.74	0.88	0.98	1.08	1.16
(Comparison with col. 1 = 100)	(100)	(140)	(166)	(185)	(203)	(219)

[1] The selection intensity value can be read from fig. 18.
[2] Ewes joined = ewes joined with the ram and given opportunity to mate.

Note that as the fertility increases then the number of animals available for selection increases. As the number needed for replacement is constant, then the percentage selected decreases and the percentage culled increases. The values for intensity can be read from the curves presented in fig. 18. The intensity clearly increases with increased lambing percentage. Using 80% lambing as a base of 100 the intensity for 180% lambing is 219. (Redrawn from Falconer[16].)

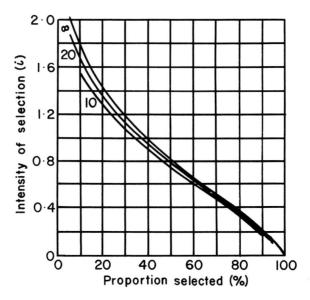

Fig. 18 Intensity of selection (i) in relation to the proportion selected. The lines are for populations of 10, 20 and an infinite (∞) population

THE GENERATION INTERVAL
This is the time-interval between generations and is defined as the average age of the parents when their offspring are born. Obviously it varies greatly between species and particular breeding plans but some general average values, in years, are as follows:

Man	25	Sheep	3–4
Horse	9–13	Dog	3–4
Beef cow	$4\frac{1}{2}$–5	Pig	2–$2\frac{1}{2}$
Dairy cow	4–5	Chicken	1–$1\frac{1}{2}$

To ensure rapid progress the aim is to keep the generation interval short. This, however, is partly limited by puberty (sexual maturity) which is the youngest age at which an animal can be bred. The generation interval can also be severely restricted by how long the breeder has to wait until sufficient data are available from an animal on which to make a decision. An example of this is waiting for completed first-lactation milk yields of daughters before a bull is widely used in a herd or through the national herd by artificial insemination.

GENETIC GAIN
The way in which the three components (heritability, selection

differential and generation interval) are put together to give an estimate of genetic gain is as follows:

Gain per generation = heritability × selection differential

$$\text{Gain per year} = \frac{\text{heritability} \times \text{selection differential}}{\text{generation interval}}$$

This equation for gain per year is the core of a breeder's programme and dictates what he gets out of it (see fig. 19).

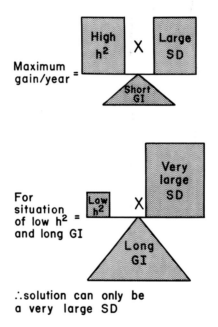

Fig. 19   **Illustrations of the components of genetic gain**

Fig. 19 shows the formula for gain per year in the form of a balance. Ideally, maximum gain is obtained from high heritability, large selection differentials and short generation interval. A typical situation in practice is where breeders have to work on traits with low heritability (e.g. reproduction) and they have long generation intervals caused by waiting for information from offspring. The only possible solution then is to achieve a very high selection differential to counter-balance the other two components.

The formula for gain per year can also be written in full as follows:

Gain per year =

$$\frac{\text{heritability} \times \text{intensity} \times \text{phenotypic standard deviation}}{\text{generation interval}}$$

or using symbols:

$$\Delta G = \frac{h^2 \times i \times \sigma_p}{GI}$$

In previous discussions about genetic gain and its components, the assumption has been made that only one trait was being considered. So it is not difficult to see how progress will be reduced if the breeder considers large numbers of characters.

Measuring genetic improvement

This is a very important issue because it is the justification or proof that all the effort put into a breeding programme has been worthwhile. Unfortunately it is difficult mainly because genetic improvement often cannot be clearly separated from environmental improvements. To do this certain standard techniques can be used as follows:

(a) CONTROLS

These are control flocks, herds or lines specifically set up by the breeder or perhaps a group of breeders, and are bred continously at random as a base or reference. The control populations will reflect only variation caused by the environment, for example seasons, disease outbreaks, changes in staff and so on and provide a genetic constant.

Control populations should be large enough to maintain effective random mating and keep the inbreeding level as low as possible (see page 89). Thus the number of males and the number of females used are critical (see Appendix I). A suggested size would be 25–50 males and 50–100 females as a guide. The main problem in control populations is 'genetic drift' – a type of random change in gene frequency. Any population can change despite efforts to avoid this. One technique to minimise this is to keep re-constituting the control population or run the control at different locations to allow comparisons between them and hence check on drift.

Examples of control populations are:

* Meat and Livestock Commission pig improvement scheme (UK) –
  16 boars and 32 gilts

* Cockle Park pig selection (UK) – 16 boars and 32 gilts
* Beltsville pig selection (USA) – 12 boars and 12 gilts
* Scottish Blackface sheep selection (UK) – 10 rams and 250 ewes.
* Cornell poultry selection (USA) – 50 cocks and 250 hens
* Hereford cattle selection (USA) – 25 cows and 1 bull

(b) REPEAT MATINGS

This is where the same parents are used in different years. It is done by using frozen semen so that progeny can be produced by the same sire (i.e. progeny of the same breeding value) but are born in different years. The difference between the progeny is due to environment, and the difference between the rest of the flock or herd and the repeated progeny is genetic gain. This may be done by using semen from a random group of sires and using it after 2-, 5- and 10-year intervals. It is a technique that is at present easier to apply in cattle than in pigs and sheep because of semen freezing problems and fertility levels.

Selection limits

In farm animals where most of the important traits are polygenic, there is little likelihood that breeders will run out of genetic variation. The limits to selection have been studied mainly in laboratory animals (insects and mice) and poultry. What happens in selection lines is that there is usually a response to selection for a while and then it slows down and eventually stops. The observed record of progress shows a definite 'plateau' as drawn for a hypothetical situation in figs. 20 and 21.

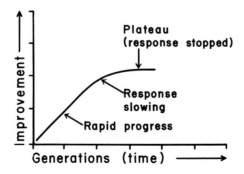

Fig. 20 **Simplified selection response reaching a plateau**

The plateau is usually caused by the population running out of usable genetic variation (fig. 20). However, if some new variation is

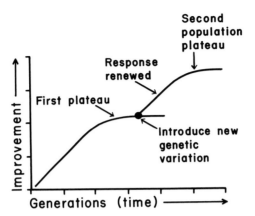

Fig. 21 **Renewed selection response reaching a plateau at a higher level**

introduced, progress from selection continues until the population reaches another plateau and so on (fig. 21). There are a number of ways in which breeders can introduce new genetic variation into their selected populations. Examples are by crossbreeding (see page 96) or by inducing mutations either naturally or artificially (see page 36). In future there is even the possibility of synthesising new variation from basic chemical components through genetic engineering.

PRESERVATION OF GENETIC VARIATION

Recently there has been a renewed interest in the preservation of genetic variation in farm livestock. This has come mainly through the desire of breeders and geneticists to store genes or 'germ plasm' (USA terminology) of the minority breeds, some of which were on the verge of extinction.

The first task has been to catalogue these breeds and this has been started by the Food and Agriculture Organisation (FAO). Germ plasm can be stored as semen for some species (the haploid state) or fertilised ova (the diploid state). In some countries the last remaining animals of a breed have been kept in zoos or game parks, and although inbreeding can cause problems in small groups of animals and stocks can be difficult to maintain through lack of fitness, there are still examples in several countries of small groups of animals being maintained successfully.

Breeding Value and the aids to selection

It was stated earlier that selection was the business of making decisions about animals in the light of 'information'. It is here that breeders have to start and consider the 'Breeding Value' (BV) of an animal. This is really its genetic worth and is what animal breeding is all about. Unfortunately, although this concept of Breeding Value has been developed for a long time, it has not been widely used in practice in all areas of farm animal improvement. Perhaps the dairy industry has used it most.

To help make decisions, there is a number of well-recognised sources from which the required information can be obtained. These are referred to in the recognised texts as 'aids to selection' and are as follows:

(a) Individual or mass selection
(b) Lifetime performance records
(c) Pedigree information
(d) Progeny performance
(e) Performance of other relatives (family selection)

INDIVIDUAL OR MASS SELECTION (PERFORMANCE TESTING)

This is used when the animal's own performance is a measure of its genetic merit. This aid to selection is used for traits of high heritability where the animal's own performance is an accurate guide as to how its progeny will breed.

The comparison of animals based on their own individual performance is usually called a 'performance test'. This term is not used so much in dairy cattle but is regularly applied to beef cattle, pigs and sheep. The theory behind it is simple. Here the best individual is selected from within a group of animals of similar age that have been similarly treated (they are contemporaries). Some practical problems may arise over what is meant by 'similarly treated' and most concern is usually over the treatment before the animals went on test (the pre-test environment). This is an important aspect and is discussed further on page 136. The point to remember is that animals should be compared within environments and not between environments whenever possible.

The real test of the value and accuracy of a performance test is whether the results (e.g. the merit-order of the animals tested, usually males) agree with the results from a subsequent progeny test (discussed on page 73). In selecting the best individuals the breeder has a *single* record of each animal's performance (the performance test), and hence an estimate of the Breeding Value (BV) for a given trait is calculated as:

BV = heritability of the trait $\times$ (individual average – average of contemporaries)

or,

BV = $h^2 \times$ (individual deviation)

LIFETIME PERFORMANCE RECORDS

Here the breeder has more than one record of an animal's performance, such as a series of lactation yields in a cow, annual fleece weights in a ewe, or repeated litter performances of a sow. A good animal (genetically) will generally perform well each season and this will be seen in above-average merit despite the ups and downs in the flock or herd average each year due to environment. The superior dairy cow, for example, will have above-average yield despite the poor quality bulk feed produced due to poor weather, whether milked by trainee or experienced staff and no matter how she is housed. If a breeder looked at one of the cow's past records he could make a fairly safe prediction about her future records. This then is the concept of 'repeatability' which is the tendency for the performance in the same animal to be similarly repeated. The greatest value of good (or high) repeatability is as a time-saver; the breeder can make a decision early in the animal's life and predict correctly what would happen over its lifetime.

Repeatability in statistical terms is the correlation between records. Repeatability and heritability are often confused – in simple terms repeatability tells how an animal will repeat a trait during its lifetime, whereas heritability tells how it will pass it on to the next generation. Repeatability, like heritability, is on the scale of 0 to 1.0 or 0 to 100%. Table 6 gives some general estimates of repeatability for traits in farm animals.

Lifetime records are extremely valuable, as animals that have produced well over a long life have proved that they may have the genetic ability to survive and then perform in their environment. They have the 'wear-and-tear' qualities so important economically in practice. Note, however, that waiting for completed lifetime records on an animal before wide exploitation will increase the generation interval, even if it increases heritability, and hence overall genetic progress will be retarded.

The breeder now has to use these repeated records to build up a Breeding Value. Here the BV is obtained by multiplying the animal's *average* deviation by a formula (sometimes called a confidence factor). Note that the animal's average deviation is obtained by taking the

Table 6 **Some general estimates of repeatability for traits in farm animals (expressed as a percentage)**

Dairy cattle	Milk yield	40–60
	Fat content	40–70
	Lactation length	20–35
	Feed conversion per lactation	50
	Milking rate	80
	Calving interval	4–20
	No. services per conception	13
	Annual non-return rate	6
	Bull ejaculate volume	70–80
Beef cattle	Birth weight	20–30
	3-month wt	42
	Weaning wt	30–55
	9-month wt	40
	Yearling wt	25
	Daily gain to weaning	18–20
	Daily gain to yearling	7–10
	Body measurements	70–90
Sheep	Ovulation rate	60–80
	Lambs born per ewe joined	15
	Lambs born/ewe lambing	30–40
	Lambs weaned/ewe joined	18–20
	Pregnancy duration	17–23
	Lamb survival to weaning	6–10
	Birth wt	30–37
	Lamb growth (daily gain)	38–48
	Fleece weight	30–40
	Wool traits (general)	50–80
Pigs	Litter size at birth	7–20
	Litter size at weaning	9–10
	Litter wt at birth	25–40
	Litter wt at 3 weeks	15
	Litter wt at 8 weeks	4–14
	Birth wt per piglet	18–40
	Weaning wt per piglet	12–15
	Adult live wt	37
	Farrowing wt	20
	Duration of oestrus	34
	Oestrous cycle length	39
	Carcass traits (eye-muscle area, lean: fat ratio)	95–98

Table 6 *(contd.)*

Poultry		
	Egg weight	80–95
	Egg shape	94
	Albumen height	74–80
	Shell thickness	66
	Shell weight	60–80
	8-wk body wt	55
	18-wk body wt	94
	58-wk body wt	88
	Sexual maturity	70
	Number of eggs	83
	Rearing mortality	72
	Laying mortality	63

deviation from the mean of each record and then averaging all deviations. The confidence factor formula (or regression equation) is:

$$\frac{kh^2}{1 + (k-1)t}$$

where k = the number of records
h^2 = the heritability of the trait
t = the repeatability of the trait

Therefore the whole formula becomes:

$$BV = \frac{kh^2}{1 + (k-1)t} \times \left(\begin{array}{l} \text{Average deviation of the dam's} \\ \text{records from her contemporaries} \end{array} \right)$$

EXAMPLE
In a beef herd selecting for weaning weight, the dams are assessed on the age-corrected weaning weight of all the calves they have produced.

Assume heritability = 0.3
Assume repeatability = 0.45

Using the confidence factor formula, a series of values can be worked out like this:

No. of records	Factor	No. of records	Factor
1	0.300	6	0.554
2	0.418	7	0.568
3	0.474	8	0.578
4	0.511	9	0.587
5	0.536	10	0.594

Thus, the more records, then the greater can the confidence be in them. This can now be built into the BV calculation using three cows A, B and C.

		Cow A	Cow B	Cow C
Deviation (kg) at				
calving	1	−20	+60	−10
	2		+24	+15
	3		+30	+20
	4			+10
Total deviation (kg)		−20	+114	+35
Mean deviation (kg)		−20	+38	+8.8
Breeding Values (kg)		0.300×(−20)	0.474×(+38)	0.511×(+8.8)
BV		−6	+18	+4.5

At this stage, the order of merit in these cows would be B, C, then A. However, the interesting questions are how will A do with some more records and will cow C really improve with her next calf. Remembering that repeatability is fairly good (0.45), it would be reasonable to assume that A would never catch up unless you knew she had suffered some environmental tragedy that was not her genetic fault (e.g. if the calf was injured and could not suckle). Cow C likewise looks as if she is going to carry on being above average in BV but is not an outstanding cow.

These comments perhaps illustrate that the Breeding Values are 'predictions' of the genetic worth of the animal and the term 'predicted breeding value' is often used. It is here that an appreciation of the difference between permanent and temporary environmental effects is important. For example, a dairy cow losing a teat by accident is a permanent effect, but a temporary effect occurs when a cow calves late or out of season and hence is temporarily penalised. This is why heritability appears to increase as the number of records increases – the temporary environmental variation is reduced. Note that averaging several records is of greatest advantage when repeatabilities are low.

PEDIGREE INFORMATION
A pedigree is simply a record of ancestry and most problems in practice arise over what value this record is. If, for example, it is only an officially-registered name and number by a breed association, then it can be of very limited value. If on the other hand it is complete with performance data as well as the pedigree names, then it is very useful. It

must be remembered that, when pedigrees were written down by the first improvers, an animal's name was all that was necessary as everyone knew how it performed. This may still be the situation in the sheep-dog world where the names of certain dogs are internationally famous because they have won many world-recognised trials.

A British breeder of sheep and beef cattle, Mr O. H. Colburn, has stressed that master breeders still use the technique of memorising in detail the pedigrees and physical characteristics of ancestors, and use the information to predict those that would transmit important traits to future generations. 'Master breeders' could be defined as successful pedigree breeders, but as Mr Colburn pointed out there are also breeders who use the same techniques but are not successful. There is also the problem that many important traits in farm animals, such as qualities of sheep dogs, draught horses and so on, cannot readily be measured to put in a 'performance pedigree' so it is easiest merely to record how many show prizes they have won.

The bases of the pedigrees are 'relationships' and these are classed as either (a) direct or (b) collateral i.e. descended from the same ancestor. The main concern of the breeder in using pedigrees is to decide how much consideration or weight to give to each ancestor because if he is using an extended or fully written-out pedigree it will show very many ancestors. The points to remember in a pedigree (fig. 22) are these:

* The animal whose parentage is recorded is called the 'subject' of the pedigree.
* Each animal in the pedigree gets half its genetic make-up from its parents – no matter at what stage in the pedigree this is examined.
* Grandparents each contribute one quarter of the genetic make-up of the subject, but the contribution made by an earlier ancestor will be even smaller. For example, an animal may have an outstanding great-grandparent, whose contribution will be one-eighth, but it must be remembered that seven-eighths come from all the others, who could have been doubtful performers.
* The accuracy of the ancestors' performance, if known, may not be highly reliable because they have been recorded under different environmental conditions.

There are still some unfounded beliefs associated with pedigree breeding. Some pedigree thoroughbred race-horse breeders and trainers believe that a much greater genetic contribution comes from the dam's side than the sire's side. Also some dog breeders believe that a sire contributes size to his offspring and the dam contributes colour. There

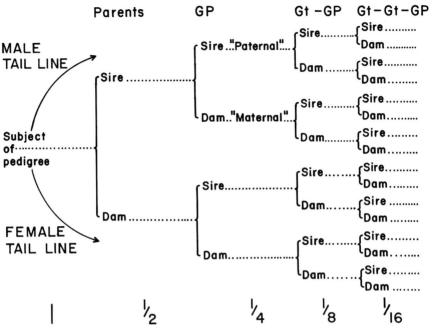

Fig. 22 **A pedigree extended for four generations**

was also the unfounded confusion expressed by Galton's first 'Law' in which he proposed a 'blended inheritance' where all the ancestors' contributions were blended to make up a total of 1.0 instead of adding up to 1.0 in each generation. According to Galton:

Each parent contributes $\frac{1}{4}$ Total $= (2 \times \frac{1}{4}) = \frac{1}{2}$

Each grandparent contributes $\frac{1}{16}$ Total $= (4 \times \frac{1}{16}) = \frac{1}{4}$

Each great grandparent contributes $\frac{1}{64}$ Total $= (8 \times \frac{1}{64}) = \frac{1}{8}$

Each g.g. grandparent contributes $\frac{1}{256}$ Total $= (16 \times \frac{1}{256}) = \frac{1}{16}$

so that $\frac{1}{2} + \frac{1}{4} + \frac{1}{8} + \frac{1}{16}$ etc. .... $= 1.0$

This 'Law' has been disproved, however. It is not accepted nowadays and should be ignored.

The business of naming animals tends to perpetuate misunderstandings over pedigrees. Breeders may use a basic name – perhaps the name they gave to their first foundation female such as 'Buttercup', a cow, or 'Belinda', a sow. Then all the subsequent offspring that Buttercup or Belinda have, regardless of the number of different sires they are mated to over their lifetime, are called Buttercup 1st, 2nd, ... up to 105th and so on. The same happens with Belinda so these become the Buttercup and Belinda families based entirely on the female line. The term 'blood line' may also be used instead of family. This can be very confusing if not clearly defined (see page 73 for discussion of family selection).

For traits of high heritability little is gained from considering ancestors and most progress can be made by evaluating the animal itself. Generally, collaterals (half-sibs or full sibs) provide more accurate data than do ancestors. Note that using ancestors is like progeny testing in reverse and presents the hazard that the breeder inevitably has to look at *selected* ancestors and not an unbiased random sample.

Professor Lush[9] summed up the emphasis given to an ancestor as depending on:

* How close is the relationship between an ancestor and the subject of the pedigree.
* How accurate are the data on the intervening ancestors if they are known.
* The heritability of the trait.
* The environmental association (correlation) between the ancestor and the subject, and between different ancestors.

The main danger in pedigree selection is that the harm done by lowering the intensity of *individual* selection is greater than the good done by making the selection more accurate. Note especially that rarely do pedigrees record the presence of recessive genes or defective animals – these animals are simply not registered. It must be appreciated that breed associations value an official pedigree as a guarantee of breed purity.

The Breeding Value concept can be used with pedigrees where the principle is to predict a BV for the subject animal in the pedigree and this is done by a statistical technique of using a regression equation. Here:

$$BV(son) = \frac{\frac{1}{2} k h^2}{1 + (k-1)t} \times \left(\begin{array}{l}\text{Average deviation of the dam's} \\ \text{records from her contemporaries}\end{array}\right)$$

where k = the number or records
     h^2 = the heritability of the trait
     t = the repeatability of the trait.

Thus for a trait like fertility (lambs born) in sheep with a heritability of $h^2 = 0.10$ and repeatability of $t = 0.15$ the values of the equation are given in table 7.

Table 7 Confidence factors (regression coefficients) used to predict a son's Breeding Value from the average of his dam's records above her contemporaries

No. records	Regression coefficient (confidence factor)
1	0.05
2	0.09
3	0.12
4	0.14
5	0.16

Again it is obvious that as the number of records increases then so does the confidence in them. Thus for a dam with three lambing records and an average deviation of 0.2 lamb above her contemporaries (corrected for age, birth rank, age of dam and run in the same environment), the BV of her son would be:

$$BV(son) = 0.12 \times 0.2 \text{ lamb} = 0.024 \text{ lamb.}$$

Note that the confidence factor formula is half that used for predicting a dam's own BV from the average of her records (on page 67) because a dam only passes on half of her genes to her offspring. If the grandam has records, they can be used as can those of the great-grandams but their confidence factors are reduced by half. In table 8, they are presented along with the dam's factors from table 7.

Table 8 Confidence factors (regression coefficients) used to predict a son's Breeding Value from the average of the dam's, grandam's and great-grandam's records above their contemporaries

No. records	Confidence factors		
	Dam	Grandam	Great-grandam
1	0.05	0.025	0.017
2	0.09	0.045	0.023
3	0.12	0.060	0.030
4	0.14	0.070	0.035
5	0.16	0.080	0.040

Clearly, it is not worth going back in the pedigree beyond the grandam no matter how many records there may be on the great-grandam.

This development of Breeding Values can then proceed to build into them the dam's information plus both the paternal and the maternal grandam's information. See Turner and Young[17] for these formulae and further discussion.

PROGENY PERFORMANCE (PROGENY TESTING)

Basing decisions on the performance of an animal's progeny is called progeny testing. It is a technique generally used for males because they are responsible for more progeny in their lifetime than any one female. Progeny testing is used in these situations:

* For weakly inherited traits
* For traits expressed in one sex (e.g. milk production)
* For traits expressed after slaughter (e.g. carcass composition)

The genetic principle behind progeny testing is simple. As each offspring represents a sample of the genes of each parent (drawn at random), then the more samples that are examined the more accurate is the assessment of the parents. Calculation of how many offspring are needed to show a real difference (and not a chance one) between sires is important for both genetic and economic reasons. Progeny testing takes time and the keeping of progeny groups for long periods can be an expensive operation.

The main points concerned with getting the best results from progeny testing are these:

* Test as many sires as possible (5 to 10 would be minimal).
* Make sure the dams are all randomised to each sire, within age-groups if possible (see page 135 for details).
* Produce as many progeny per sire as is possible (aim for at least 10–15 of either sex per sire for growth traits but up to 300–400 offspring may be needed for traits like calving difficulty and fertility).
* No progeny should be culled until the end of the test.

PERFORMANCE OF OTHER RELATIVES
(FAMILY SELECTION)

The term 'family selection' is generally applied to situations where relatives are used to help make decisions. In practice there is generally a great deal of confusion over family selection because breeders use different definitions of the term 'family'. The whole business of 'families and blood lines' is part of the tradition of pedigree breeding.

Families can be broadly classified into three types:

(i) SIRE FAMILIES
These are progeny of one sire:

* Out of different dams – born in the same year (contemporaries).
* Out of different dams – born over a number of years.

(ii) DAM FAMILIES
These are progeny of one dam:

* By different sires – born in the same year as can be done by superovulation of the dam before artificial insemination with mixed semen from a number of sires. The progeny may be identified to sire by blood typing.
* By different sires – born over a number of years.

(iii) SIRE AND DAM FAMILIES
These are progeny by one sire out of one dam. Again by embryo transfer a number of offspring can be obtained as contemporaries born in the same year, or offspring can be obtained over a number of years.

It is obvious how the carrying on of family names in pedigrees (discussed earlier) adds to the confusion. Professor Falconer's explanation[16] of family selection is presented in fig. 23 where he used hypothetical families.

Here there are four systems of family selection A, B, C and D. In each system there are 25 individuals divided into five families (five per family). The mean of each family is marked by a cross. A, B and C are identical arrangements and ten animals have to be selected from each system shown by solid circles.

A. This is *individual selection* – the best ten individuals are kept regardless of the family means. Note that none were kept from family 3, only one was kept from family 4 and four were kept from family 5.
B. This is *between-family selection* where only the animals from the best families (2 and 5 with highest means) are kept.

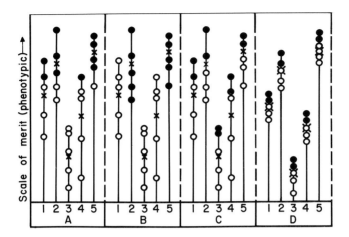

Fig. 23   **Different methods of family selection (redrawn from Falconer)**

C. This is *within-family selection* where the best animals from *each* family are selected.
D. This is a situation where within-family selection is most useful i.e. where differences *between* family means are large and the variation *within* families is small.

It is important to note the difference between the within-family and the between-family methods of selection. If you select within a family then every family is represented in the next generation. If you select between families, then only some families are represented and there could be a rapid build-up in the rate of inbreeding.

Genetic theory stresses that family selection is most effective when the *genetic* relationship between members of the same family is large, and the observed (phenotypic) relationship between members is small. This is a situation found in families produced after inbreeding (inbred lines). The breeder's problem is to make a decision based on the animal's own traits, its deviation from the family average and how good that particular animal is compared to others – provided he knows how the family was defined.

COMBINATIONS OF SELECTION AIDS

Breeders often use combinations of these various selection aids and most common would be the combined use of individual and family selection, the decision depending mainly on the size of the heritability of the traits. Where heritability is low the use of family data is most valuable as it reduces the chances of making the wrong decisions. Family selection and progeny testing are different aspects of the same thing.

SELECTION METHODS

Once the breeder has decided on the information he is going to use to aid his selection then he actually has to do the selection. There are three methods of selection; tandem selection, independent culling levels and index selection.

(i) TANDEM SELECTION

This is where the breeder selects for and improves one character until it reaches an acceptable level, and then he leaves it while he selects for another and so on for a third. This is illustrated in fig. 24. Note that trait

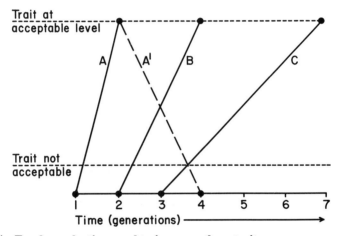

Fig. 24   **Tandem selection used to improve three traits**

A was improved quickly in one generation whereas B took more time (two generations) and C took very much longer to improve (four generations). Note also that this is a situation where A remained stable when he worked on B, and both A and B remained where they were when he worked on C. In other words the traits are assumed to be independent. If they were not, then the situation could be as seen by the dotted line A' where as B went up A came down, i.e. a 'see-saw effect' caused by a genetic antagonism between them. This is the genetic correlation discussed on page 83.

This is probably the most commonly used technique in practice. It is seen regularly where for example a dairy breeder buys a bull to bring up the fat level in his herd for a while, then uses sires noted for yield; a pig breeder buys a boar to increase length then he will continue selection for growth; a sheep breeder aims to select for fertility but as the wool price rises he goes for fleece weight. If wool prices collapse and lamb prices improve he will swing to sires with good weaning and hogget weights. It seems to be a 'stop-go' policy but often meets the needs in rapidly changing economic conditions.

(ii) INDEPENDENT CULLING LEVELS

Here accepted levels of performance are set and any performance that fails to reach these levels means automatic culling. It is like an examination system with different pass marks for each subject but if the student fails one subject then he fails the lot. There is no compensation for poor performance in one trait by brilliance in another. This method is most useful when traits are reduced to a minimum and where culling is done at different stages in an animal's life. Some examples in sheep would be:

* *Lambs:* cull lightly at weaning for severe structural faults only.
* *Hoggets (yearlings):* cull on live weight and/or fleece weight.
* *2-tooths (18 months-old):* cull on reproduction (reared a lamb or not).
* *Older ewes:* cull on reproduction (reared a lamb or not).

Here there are different selection pressures at each stage.

The culling levels should be decided with a knowledge of heritability, the overall economic importance of the trait and the number of animals available for selection. In practice it is easy to get 'culling on eye-ball' out of balance with the 'culling on records', with the result that the breeder culls too deeply too soon. On the other hand where no records are available a breeder may have to reduce numbers, for example before a period of food shortage, and a culling level is thus set by necessity and not by choice. It is the old problem of balance in decisions – difficult to achieve and even more difficult to keep consistent over a period of years.

(iii) INDEX SELECTION

This is where the individual specifications for a number of different traits that can vary greatly are combined into one value for the animal – called a total score or an index. Here the high merit in one trait can certainly be used to make up for the deficiencies in others. An index is simply a means of putting a whole lot of different information into one value. However, the most exciting thing about an index is that it is not just a means of summarising the historical records of the animal – it can go a stage further and predict the future genetic and financial worth of the animal i.e. its breeding value in financial terms. To do this requires a considerable amount of data such as:

* The variation seen in each trait – the phenotypic standard deviation.
* The heritability of the traits.
* The phenotypic relationships (correlations) between the traits.
* The genetic relationships (correlations) between the traits.
* The Relative Economic Value (REV) of the traits.

The REV often causes confusion among users of an index because it is not based on the actual prices in use at one particular time, but rather the *relationship* between the prices of the components over a period of time. Thus the relation between prices in the past is used to predict their relationship in the future.

The aim in computing an index is to derive an estimate in which the various traits are appropriately weighted to give the best prediction of the animal's breeding value i.e. what it will produce when it breeds. An important aspect of an index is that if one component is missing then benefit can be obtained by predicting the missing one from the others that are present. So basically it is a large weighting exercise that in the past has been beyond the means of practical farmers, but with the rapid increase in the use of computers there is no reason why future breeders could not use index selection more extensively – even if it was to modify generalised indexes to suit their individual circumstances. It is important to ensure, however, that the technical sophistication achieved is economically rewarding.

It often seems that the main problem with a selection index is that scientists cannot explain to breeders how it works. Some breeders have developed their own 'home-spun' indexes in an attempt to grasp some of the benefits but avoid the complexity. However, care is needed with indexes to find out how they were calculated before they are accepted.

In a selection index used for sheep (New Zealand National Flock Recording Scheme, *Sheeplan*), the main features are shown in a simplified table (table 9). Column (1) is a predicted average genetic gain in the four characters while column (2) is the relative economic value among the traits. Note that NLB is very important followed by wool production (HFW). Body weight (HLW) is taken as zero as bigger

Table 9 **Features of a selection index for sheep***

Trait	Predicted average genetic gain	Relative Economic Value (REV)	Contribution to economic response (percentage)
	(1)	(2)	(3)
No. lambs born (NLB)	0.05 lambs	554	65
Weaning wt (WWT)	0.5 kg	24	28
Hogget live wt (HLW)	1.2 kg	0	0
Hogget fleece wt (HFW)	0.03 kg	92	7

*Calculations based on three lambings per dam and a selection differential of one standard deviation.

animals eat more so the net economic gain will be zero. Column (3) shows the contribution of each trait to economic response by using the index. The author has developed a diagram to try and explain in simple pictorial terms how the index works. This is shown in fig. 25.

Fig. 25 shows the four components of the index drawn as separate sheep, using the abbreviations in table 9. Heritabilities are drawn on the sheep for each trait as a number on the scale 0 to 100. The genetic correlations are shown above the sheep and the phenotypic ones below, again as numbers out of 100. The term 'physical links' may be used to describe the phenotypic correlations or 'genetic links' to describe the genetic correlations if correlations are not understood. These correlations can be added to the picture of the sheep in a series as overlay transparencies for an overhead projector along with the lines added to show the Relative Economic Value (REV) of the traits in the combined index for a ram. The HLW does not contribute directly to the index but is important as an indirect aid to improving NLB through the genetic correlation between HLW and NLB (of 20).

Genetic theory says that an index is predicted to be $\sqrt{n}$ times as efficient as independent culling levels when n = the number of traits involved. The greater the number of traits involved, the more reliable the index becomes, for example:

$$2 \text{ traits}: \sqrt{2} = 1.41$$
$$10 \text{ traits}: \sqrt{10} = 3.16$$
$$16 \text{ traits}: \sqrt{16} = 4.0$$

Poultry breeders have used indexes with up to 16 items for egg production alone, some traits appearing in the index in various forms

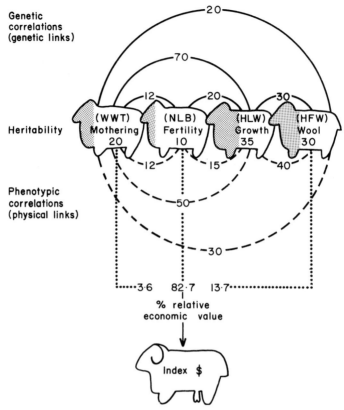

Fig. 25 **A diagrammatic explanation of a selection index**

such as in individual records and various family averages. Pig breeders have used indexes for some time and these have included up to 9 traits. An example was that used in the British meat and Livestock Commission's 'on-farm' performance test.

$$Y = 101.2W/A - 0.892(C+K) + 0.0202W + 2.6$$

where

    Y is index score of breeding value
    W is live weight (kg) on day of test
    A is age (days) on day of test
    C and K are fat depths (mm)

The index was designed to select for efficiency of production of lean meat and was constructed by the MLC from parameters estimated from data collected in the central testing stations. Sheep and beef breeders have been much slower to use index selection but this is changing in some countries.

RELATIONSHIPS BETWEEN TRAITS

Breeders readily recognise that some characters, or aspects of them, affect others. This is part of the whole complex of animals becoming adapted to their environment and the term 'fitness' is often used to describe this. Breeders are aware for example that increasing milk yield affects a quality such as fat percentage, and increasing live-weight gain may increase carcass fat deposition; or increasing body weight could increase fertility and thus increase mortality at birth. The list of examples is endless.

Relationships in statistical terms are expressed by correlations. These are calculated as values on a scale from –1.0 through 0 to +1.0, and show how one factor (often called a variable) changes as another variable changes. Positive correlations show that as one trait goes up then so does the other, while negative correlations show that as one trait goes up then the other goes down. It is possible to get some idea of the correlation between two traits by plotting them on a scatter diagram and drawing points where the values for each animal meet. Fig. 26 shows the types of scatter that are associated with different correlations between some traits in livestock.

Correlations are broadly classified as follows:

–1.0 to –0.6 = high negative
–0.5 to –0.4 = medium negative
–0.3 to –0.2 = low negative
–0.1 to +0.1 = negligible (zero)
+0.2 to +0.3 = low positive
+0.4 to +0.5 = medium positive
+0.6 to +1.0 = high positive

For a complete and detailed explanation of correlation and regression, the reader should consult books on biometry and statistics.[18, 19]

TRADITIONAL RELATIONSHIPS BETWEEN TRAITS

Breeders work amidst an enormous legacy of belief that has been passed down as part of traditional teaching. This is guaranteed to generate heated discussion between scientists (who tend to disbelieve) and breeders (who tend to be believers). Here are some examples:

* Flat bone denotes good meat potential.
* Certain coloured fibres in a sheep's fleece denote hardiness (e.g. red kemp in Welsh Mountain sheep).
* Good milk veins on a cow's belly denote high milk yield.
* A thin skin (when pinched between finger and thumb) denotes good milk potential in a dairy cow.

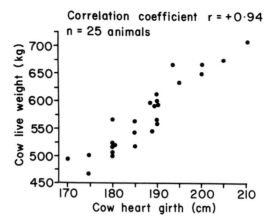

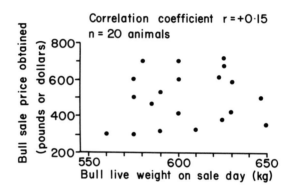

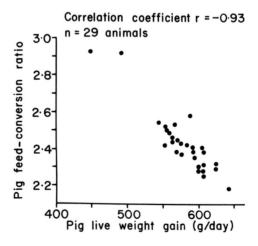

Fig. 26   **Different types of relationships (correlations) between traits**

* Good heart-girth in a cow denotes good constitution (an ability to thrive and withstand stress).
* A sheepdog pup with a black palate will be a good worker.
* A kind eye (friendly look) in most stock denotes good temperament and hence good performance.
* A feminine head, as opposed to a masculine head and thick neck, denotes good lactation performance and fertility in a cow.

The list could be endless. Some of the relationships are physiological nonsense and must be ignored, but there could be some truth in others so care is needed until concrete evidence is available. It is interesting to note that some of these firmly-held beliefs may be specific to a breed, a district or a country, and may be almost part of the social culture.

It is easy to understand how these relationships developed because for important traits like thrift, constitution, intelligence that were difficult to measure, the breeders had to find some easy-to-see trait as an indication of merit to be able to make a decision at all. Generally no harm would be seen to be done and faith in, and respect for, the 'master breeder' who propounded it probably gave the relationship a sort of blessing. So despite our sophisticated modern age, breeders still have these traditional beliefs – and the disagreements between scientists and breeders look as if they will continue for a long time yet.

DIFFERENT TYPES OF CORRELATIONS
In animal breeding it is important to recognise three different correlations and to describe them we need to go back to the basic equation of:

$$P = G + E$$

i.e. Phenotype = Genotype + Environment

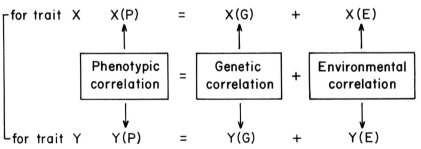

This shows that there are correlations between the phenotypes, the genotypes, and the environments of these two traits in the animal. The above equations highlight the fact that if the two traits *appear* to be related (i.e. there is a phenotypic correlation between them), then this could be for two reasons:

(a) Some of the *genes* affecting one trait also affect the other i.e. the genetic correlation between X(G) and Y(G). This is pleiotropy, discussed earlier on page 37.

(b) Some non-genetic or *environmental* factors affecting one trait also affect the other, i.e. the environmental correlation between X(E) and Y(E). An example would be in a starved flock of ewes where all those that bore twins would have finer wool. This is called 'hunger fine' wool as the ewe does not have enough feed energy to divert to wool and reproduction has higher priority. Clearly this is an environmental problem and not a genetic relationship between fine wool and twinning.

Genetic correlations are of greatest interest to breeders for three main reasons:

(i) They can indicate how things are likely to change in the next generation. Thus selecting, say, for growth rate in this generation by picking the heaviest bulls at 400 days of age, the birth weight of their calves (the next generation) will also increase and could result in calving difficulty.

(ii) They can be used to plan counter-selection measures to prevent any correlated changes that are not wanted. Thus in the above example where birth weights were increased by increasing 400-day weight, the breeder should look for individual sires (by progeny testing) whose calves were produced without difficulty despite their being heavier at birth.

(iii) They can be used in situations where one trait may be difficult to improve and another correlated trait can be used to help improve it. The 'difficult' trait may be difficult because of practical problems in measuring it accurately or because it has low heritability. An example would be where a pig breeder wanted to improve feed efficiency but did not want to record the feed eaten. He would simply select for gain and would know, from the good genetic correlation (negative), that as gain increased feed conversion became better, i.e. less kg feed/kg gain.

Another example, in sheep, would be that as fertility has low heritability, direct selection would bring slow improvement. This could be assisted by selecting for yearling body weight that is both

Table 10 **Some genetic correlations (above diagonal) and phenotypic correlations (below diagonal) for traits in farm animals**

Dairy cattle		LY	FC	SNF	TP	C	L
Lactation yield	LY		−0.35				
Fat content %	FC	−0.25		0.50	0.57		0.37
Solids-not-fat %	SNF		0.55				
Total protein %	TP		0.57	0.94			0.41
Casein %	C		0.56	0.82	0.96		
Lactose %	L		0.37	0.67	0.41	0.41	

Beef cattle		BW	WW	BWG	FW	YW	CMA
Birth wt	BW		0.58	0.38	0.64		−0.11
Weaning wt	WW	0.39		0.98	0.79	0.67	
Birth-wean gain	BWG	0.23	0.97		0.73		
Final feedlot wt	FW	0.42	0.65	0.63			
Yearling wt (pasture)	YW		0.64				
Cow maternal ability	CMA						

Sheep			NLB	WWT	ALW	WLW	SLW	HFW
No. lambs born	NLB			0.12	0.15	0.17	0.20	−0.05
Weaning wt	WWT	(4 m)	0.12		0.80	0.75	0.70	0.20
Autumn live wt	ALW	(7 m)	0.13	0.70				0.30
Winter live wt	WLW	(10 m)	0.14	0.60				0.30
Spring live wt	SLW	(14 m)	0.15	0.50				0.30
Hogget fleece wt	HFW	(14 m)	0	0.30	0.35	0.40	0.40	

Pigs		DG	FE	KO	CL	BT	EM
Daily gain	DG		−0.76	−0.19	0.14	−0.15	−0.11
Feed efficiency	FE	−0.73		0.01	−0.08	0.21	−0.34
Killing out %	KO	−0.17	−0.05		−0.40	0.28	0.36
Carcass length	CL	0.07	−0.04	−0.19		−0.30	−0.08
Backfat thickness	BT	−0.07	0.19	0.19	−0.22		−0.28
Eye muscle area	EM	−0.03	−0.16	0.15	−0.05	−0.13	

Poultry		AFE	HDP	EW	BW	V	SG
Age at first egg	AFE		−0.4	0.4	0.4	0.7	−0.1
Hen day production	HDP	−0.3		−0.5	−0.1	0.4	0.1
Egg wt	EW	0.2	0.2		0.25	−0.2	0
Mature body wt	BW	−0.2	−0.1	0.3		0	0
Viability	V						−0.2
Specific gravity	SG						

phenotypically and genetically correlated with fertility (lambs born). This is called indirect selection.

This whole subject is dealt with in the main text books[16] as a study of the correlated response to selection and the relative merits of both direct and indirect selection.

A practical view is that the stud breeder should obviously concentrate on the genetic parameters (estimates) like heritability and genetic correlations. However, commercial breeders who will be concentrating most of their selection on females and will be buying in sires are more concerned with phenotypic selection, and hence are more interested in parameters like repeatability and the phenotypic correlations between traits.

Care is needed with phenotypic correlations between traits until the size and sign of the genetic correlation between them is known. The greatest hazard is where there could be a positive phenotypic correlation masking a negative genetic one, and where real progress would mean going in reverse. Phenotypic and genetic correlations between traits as well as heritabilities are usually presented in tables, and typical examples of these are shown in table 10. The genetic correlations are shown above the diagonal, with the phenotypic ones below.

Breeding and the environment

This is a very important part of the study of applied animal breeding and a great deal has been written about it over the years by scientists and breeders alike. Despite this there still seems to remain some confusion. The confusion seems to arise through a misunderstanding of the effect that the environment has on the animal (the phenotype), and the effect it has on its genes (the genotype). The environment can have a direct effect on the phenotype, for example, through nutrition, disease incidence and management, but can have no effect on the genotype. The genotype can only be indirectly affected by the environment by the alteration of gene frequency so that certain types are selected as parents for the next generation and others are ignored.

The problems that concern breeders can be best expressed as a series of questions that are regularly asked on this subject. Examples are these:

* Will the 'hardiness' of sheep for mountain and hill conditions be adversely affected if ram-breeding flocks that supply them are located on good lowland conditions?

* Will selection of beef bulls by performance testing for growth under intensive concentrate-fed conditions also identify sires whose progeny will grow well at pasture?
* Will selection for high egg production in poultry kept in cages produce birds that will lay well on deep litter or free range?
* Will pigs selected for growth and carcass merit on wet feeding regimes produce progeny who are also superior on dry feeding regimes?
* Will progeny testing of dairy bulls for high milk yield using progeny fed at pasture and milked in large herds in rotary cow-sheds, also produce progeny that do well when milked in cowsheds in small herds under intensive feeding conditions?
* Will selection for growth rate in cattle in temperate climates produce offspring that perform well in the tropics?

The questions are limitless and complex and the whole subject is really concerned with what is called genotype–environment interaction or G×E. The question is to see whether a genotype responds differently (interacts) in different environments. A hypothetical example would be where two cattle breeds were tested for good growth rate in two contrasting environments like this:

	Temperate climate	Tropical climate
Hereford (B taurus)	Good	Bad
Brahman (B indicus)	Good	Good

Here the Hereford has failed to perform in the tropics (because of heat stress) whereas the Brahman has grown well in both environments.

To try and answer some of these questions that concern breeders, geneticists have worked mainly with laboratory animals to try and elucidate the basic principles. The general simplified conclusion from a great deal of work is that for genetic reasons it is best to select and breed animals in the environments in which they have to perform. This is also the easiest thing to do for practical reasons as the breeder does not have to worry about selecting for things in one environment that are needed in another. However, it is important to stress the need both to select and to breed in the same environment. It is when these are split that problems can arise. This would happen where a flock breeding rams for the mountain is kept entirely on the lowland, with no ewes ever coming down from the mountain to supplement the ram-breeding flock.

In the previous questions, for example in the bull performance test, difficulties could only arise if selection on feedlot produced stock that could not walk and hence graze pasture themselves. Similarly, in poultry, if selection in cages altered the bird's physiology or behaviour so that it could not lay on litter or at range, then there would be serious problems.

The greatest concern for breeders is where there are many genotypes (e.g. breeds) to test and many environments. Among all the possible combinations of breeds and environments there are usually some important interactions. In general, research results would show up to the present that $G \times E$ is not very important in dairy and beef cattle and is more important in sheep, pigs and poultry. Although it may appear that poultry are kept in controlled environments, there are still important variables like the number of birds per cage, stocking intensity, cages versus litter, feeding, lighting and temperature regimes, exposure to certain diseases and so on, any of which could interact with specific genotypes. It is obvious that breeders have to cover themselves against the most important variables, although they cannot cover all the hazards of commercial environments.

This subject becomes of special interest when the needs of breeders around the world are considered. With the current ease of transferring genotypes by semen or embryos, the responsibilities of breeders are greater, especially if minority breeds in one area become internationally popular. An example would be the world popularity of Charolais cattle, which were bred in small herds in France, but whose progeny are now performing on sugar by-products in Cuba and Brazil, on dry pastures in the Australian outback, on steep hill country in New Zealand and under heat stress in Fiji. Breeders in each country will eventually develop their own strains but breeders returned to France in the initial stages for new blood lines. The French breeders thus had a big genetic responsibility which could have involved consideration of $G \times E$ interaction on a world basis.

MATERNAL ENVIRONMENT
Breeders are very much concerned with the effects of the maternal environment – sometimes referred to as the 'maternal handicap'. This is the environment that a mother provides for her offspring from

conception to birth and then up to weaning. It affects the offspring's phenotype and not its genotype. The genotype of an animal cannot be affected in this way as was formerly believed in the theory of *telegony*. This was a belief that the effects of previous pregnancies could affect later ones, for example when a pedigree bitch was mated to a mongrel dog it was thought that this would affect subsequent litters by pedigree dogs. This theory, however, has been proved false.

Another erroneous belief was that the physical environment of the dam could affect its offspring's genotype, for example schooling a mare during pregnancy would increase the chances of the foal being a good jumper. This is also unacceptable.

It is most important to be aware of situations where improved performance in an offspring may appear to be due to superior genotype but is in fact caused by the maternal environment. This is usually seen in cross-breeding as in the classical example of the large Shire crossed with the small Shetland horse. The F_1 offspring out of the Shire mare (by the Shetland stallion) were three times heavier at birth and one and a half times heavier at four years old than the offspring out of the Shetland mare (by the Shire stallion). Similar work was repeated with the large South Devon and the small Dexter breeds of cattle.[3] The problem of maternal environment is especially important where litters of offspring are involved. Here litter size and hence competition in utero before birth can limit subsequent genetic expression of growth traits. (See Learner and Donald[1] for further discussion.)

Breeding methods

So far, the discussion has concerned how the breeder selects parents for the next generation. His next task is to decide how to breed them, i.e. how to mate them together. This is the area of 'breeding methods' and is classified in table 11 where the methods are first divided broadly into *closebreeding* which is the mating of related parents, and *outbreeding* which is the mating of unrelated parents.

INBREEDING

Inbreeding is the mating of animals that are more closely related to each other than the average of the population i.e. mating animals that have one or more ancestors in common. So the key to searching a pedigree for evidence of inbreeding is to look for those 'common ancestors' that appear on both sides of the pedigree. If the parents of an animal (the subject of the pedigree) have common ancestors close up in the pedigree, then the

Table 11 **Classification of different breeding methods**

Close breeding *(mating relatives)*	*Outbreeding* *(mating non-relatives)*
Inbreeding	Crossbreeding
Line breeding	Outcrossing
	Backcrossing
	Topcrossing
	Grading up
	Mating likes
	Mating unlikes

offspring will be inbred and this degree of inbreeding can be calculated and expressed as the 'inbreeding coefficient'. The level of inbreeding thus depends on the closeness of the relationship between the parents. Either or both parents may be inbred themselves, but if they are not related to each other then the subject cannot be inbred.

The 'inbreeding coefficient' is the rate at which heterozygosity is reduced (or homozygosity is increased) per generation in the population. The calculation of an inbreeding coefficient is described in Appendix I.

Inbreeding simply reduces the number of gene pairs that are heterozygous in the population and increases the proportion of gene pairs that are homozygous, regardless of whether they are good or bad. The main value of inbreeding is to concentrate genes in the population and to retain known merit by following a particular animal with one of its close relatives.

Inbreeding is often used because of necessity rather than choice, especially by 'top' breeders whose stock have reached a high level of performance. They may need a new senior stud sire and cannot find one in anyone else's herd that is as good or better than their own. In such a situation the best possibility would be a relative, perhaps a half brother or son of the sire they are using.

INBREEDING FEARS

No other aspect of animal breeding seems to have as much fear and mystery built around it as inbreeding. It seems as though some of this fear has come from human genetics and religions with the fear of increasing consanguinity (the same blood relationship). The fears of inbreeding in animals are about 'inbreeding depression' which results in a reduced performance or fitness of the animal for its job. Some traits are affected more than others. Initially inbreeding may throw up odd defects such as undershot jaw, dwarfism, odd colours and so on, but these are usually of limited economic significance in the early stages and are basically recessive genes being thrown up. They can usually be reduced in a reasonable-sized flock or herd by culling the carriers, especially the males, if the breeder cannot afford to cull the female carriers. It would then certainly pay to go to a completely unrelated sire that was guaranteed free from the defect but these may not always be easy to find.

Known 'carriers' of these defects can be used to mate with other animals for testing. An example of this would be the mating of a bull which was intended for widespread AI use to about 15 to 20 of his own daughters. If there were any recessive genetic defects in the stock then they should appear in these matings. Then the breeder has to assess how important these defects are: there seems little sense, for example, in slaughtering an outstanding Holstein bull that will improve both milk and fat yield because he throws one red calf in every 10 000 progeny. Somewhere the breeder has to decide what is acceptable and what is not, and this is usually a very difficult decision.

What is termed 'inbreeding depression' can be more serious than throwing up the odd recessive genes. It is the gradual lowering in performance of traits and is seen especially in characters like fertility, survival and size. The breeder may not suspect inbreeding depression and may waste time searching for the cause of the problem in the environment such as disease, poor feeding, seasonal effects and so on. Very rapid rises in inbreeding usually bring out the problems (if there are any) more quickly than a slow build-up. Table 12 shows the different build-up in inbreeding level (expressed as a percentage) with different systems of inbreeding.

The one, three and five sires are *new* sires chosen each year in a closed flock or herd, so if an average generation interval of five years is assumed, then the breeder needs 5, 15 or 25 *new* sires used per generation. Clearly, self fertilisation as used in maize causes very rapid rises in the inbreeding level, as does mating full sibs (full brother to full sister) which is the most intense rate of inbreeding in animals followed by the sire back on the offspring. However, it can be seen that by using relatively few *new* sires each year, a flock or herd can be closed for a good number of generations before the average inbreeding level builds up to high levels. Note that these values are based on random mating: if deliberate attempts were made specifically to avoid mating close

Table 12 Examples of different intensities of inbreeding expressed as a coefficient of inbreeding %

Generation	Self-fertilisation	Mating full sibs	Sire × offspring	One-sire herd	Three-sire herd	Five-sire herd
1	50	25	25	2.5	0.8	0.5
2	75	38	38	5.0	1.6	1.0
3	88	50	44	7.5	2.4	1.5
4	94	59	47	10.0	3.2	2.0
5	97	67	48	12.5	4.0	2.5

relatives in the flock or herd, the chances of inbreeding build-up would be lessened.

Inbreeding is the most powerful tool the breeder has to establish uniform families in a population that are distinct from each other. This alters gene frequency and makes the members of the same family more likely to inherit the same genes because their parents were related. The way in which inbreeding alters the structure of a population is shown in fig. 27. The diagram shows full-sib matings. In each case A, B and C, the

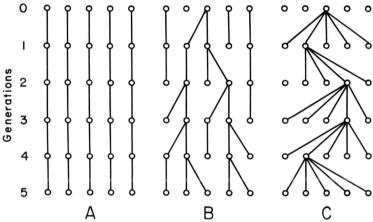

Fig. 27 **The effect of inbreeding on breed structure. (Redrawn from a diagram used by Falconer in a research paper.)**

lines start from an outbred stock made up of a number of separate families. The main points are:

* A: each family contributes to the next generation. It is a parallel line system.
* C: only one family is used to provide the parents for the next generation. This is a single-line system.
* B: is a compromise between A and C because inevitably some lines die out and rarely can enough parents be found in one family for the next generation. This is especially the case when inbreeding builds up, and fertility may be lowered as a result.

In practice the best aim would be to start off with many lines and end up with a few superior ones. A useful technique is to divide a flock or herd into groups and rotate sires around these sub-groups to slow down the build-up of inbreeding. It also means that a longer working life can be obtained from an expensive purchased sire who would normally have to be disposed of after three seasons when he would mate his own daughters. He could be used for three seasons in each sub-group if no better sires became available. However, it must be remembered that this would lengthen the generation interval.

If a breeder is forced to inbreed, the benefits he gains by more effective selection will generally counteract any deficiences caused by the slow build-up in inbreeding level. Breeders are very aware of the power of inbreeding in increasing 'pre-potency' i.e. the ability of the animal to breed stock like itself. Pre-potency will increase as the inbreeding increases.

LINEBREEDING
Linebreeding is like another form of inbreeding: it is often described as slower inbreeding in which the breeder aims for the benefits while trying to avoid the troubles. It is trying to make haste slowly, or is like a ratchet mechanism holding known benefits while slowly trying to gain more merit one notch at a time.[9] Some observers say that breeders use the term linebreeding if the results are good and call it inbreeding if the results are a disaster. Linebreeding does not seem to have the traditional fears that are associated with inbreeding.

Linebreeding is 'retrospective breeding' where for example the merit of an ancestor (now dead) has been realised, and there is urgent need to regain some of its genetic merit, the nearest of which would be that in a son or grandson. It involves a deliberate concentration on a particular ancestor. If semen had been collected from a sire and stored as an insurance there would be no problem, but then this could lead to very intense inbreeding so using a near relative could be safer.

SOME GENERAL POINTS
It is highly desirable to determine the most appropriate rate of inbreeding. The points to consider have been summarised by Lush.[9] The best rate of inbreeding depends on :

* The skills of the breeder in his selection.
* The frequency of the undesirable genes in the population.
* Any linkage between good and bad genes in the stock.
* The amount of dominance, espistasis and environmental effects that may deceive the breeder.
* The size of the population.

Professor Lush considered that 6% inbreeding was the 'stop, look and listen' stage. If the breeder has to go outside his flock or herd to avoid trouble then he should first make a mild outcross to a population that has a similar programme to his own. Otherwise, all his initial progress could be severely diluted or even lost in one outcross that went wrong. So the answer to the breeder's question whether he should inbreed would be: try to avoid it unless there is a clear reason for doing so and a plan to deal with the result whether it be successful or not. Inbreeding usually increases slowly in most flocks and herds and is reduced completely with an outcross to an unrelated sire.

SOME EXAMPLES OF INBREEDING IN ANIMALS
In poultry at Iowa (USA) after 21 years of inbreeding the coefficient was 85%. Deterioration varied in each line but egg production consistently declined with inbreeding – for example 0.43% for each 1% of inbreeding coefficient. Further poultry work at Minnesota (USA) showed that after 13 years inbreeding coefficients had reached 60 to 70%. The inbreds produced fewer eggs and suffered greater mortality than the outbreds.

In pigs in the USA a large trial across the States established more than 100 inbred lines where a 30% inbreeding coefficient was reached. Inbreeding decreased litter size at birth and viability between birth and weaning. Post-weaning growth rate was also lowered by inbreeding, and inguinal hernia, cleft palate and haemophilia had appeared in the lines. The conclusion from this American work was that it seemed possible to maintain a closed herd indefinitely as long as the inbreeding coefficient did not exceed 3 to 5% per generation and if selection was applied continuously for performance traits.

Work at Edinburgh with full-sib matings in 146 lines of pigs showed that inbreeding degeneration was seen through lowered fertility and viability. Degeneration was least in the first generatons when the pigs and not their dams were inbred. It got worse when both dams and their offspring were inbred and the most highly inbred pigs were shorter in body length and fatter than their outbred controls.

In cattle, the history of the Shorthorn showed that inbreeding coefficients were zero in 1790, 20% in 1825 and 26% in 1920. Bates kept the Duchess shorthorns around 40% inbreeding after about eight generations, which is probably a record in cattle. However, the Wild White Park cattle of Chillingham in Britain must have the highest inbreeding level of all cattle although its value is unknown. The herd is a small population (13 animals in 1947 and 50 in 1977) in which the use of different sires is restricted through the 'king bull' hierarchy. The king bull is the main one to serve the cows until he is beaten in battle by a younger bull.

Most modern breeds of dairy cattle are being inbred at about 0.5% per generation but where a limited number of popular bulls is used to bring about widespread improvement in performance, the inbreeding level can build up. A good example is in Denmark where the very efficient use of AI as a breed-improvement tool in the Danish Jersey caused concentration on a few outstanding sires and rapidly built up the inbreeding level between 1915 and 1956. Here the benefits of national breed improvement were clearly weighed against the disadvantages. Very high levels of inbreeding in cattle have been shown to reduce calf birth weight and reduce survival as well as milk and fat yield. Mature size and weight are also reduced.

In sheep most of the studies have been done on the Australian Merino and American Rambouillet.[17] Inbreeding appeared to decrease all wool traits as well as fertility.

OUTBREEDING

Outbreeding is the very opposite to close breeding: it is the mating of animals that are less closely related than the average of the population from which they came. It is the standard method of increasing variation, both phenotypic and genetic, in the population. The heterozygosity of the population is increased by outbreeding and as a result, general fitness and adaptation of the animal to its environment are usually seen. The different types of outbreeding are as follows:

CROSSBREEDING

Crossbreeding has always played a major role in livestock improvement. Most of the purebreds of today were the crossbreds of yesterday and the problem of defining the difference between a 'crossbred' and a 'mongrel' still exists. From general usage it might be assumed that a mongrel was the result of an accident, while a crossbred was the result of a planned mating. In some traditional areas of breeding, all crossbreds are assumed to be inferior to purebreds or straightbreds. In others, the F_1 cross is acceptable but all subsequent crosses appear to have mongrel status.

The first task in discussing crossbreeding is to define what is being crossed. Crosses can be made between the following: species; breeds; strains or lines; inbred lines. In crossing terminology the sire breed or individual animal is always listed first. This is important as the reciprocal cross (the opposite) may give different results.

(a) CROSSING SPECIES

This has not been widely exploited in animal production because of the technical difficulty of getting species with different numbers of chromosomes to cross. The sperm may fertilise the egg but generally embryo survival is low. If the species-cross survives to sexual maturity then it is usually sterile. However, future developments in genetic engineering could bring about changes here if there was a need to breed from the crosses. Many species crosses are mainly of zoological interest at present, for example lion ✕ tiger = liger. However, species crosses do have potential for animal production in difficult environments such as the very hot and the very cold areas of the world.

Examples for hot climates:

Ass ✕ Zebra = Asbra (Africa)
Horse ✕ Grevy's Zebra = Zebroid (USA)
Horse ✕ Ass = Mule

Examples for cold climates:

Cattle ✕ Buffalo = Beefalo (Canada and USA)
Zebu ✕ Yak (Himalayas)

Crosses between goats and sheep have always created interest, the result being sometimes called a 'geep' and viable specimens have been produced by genetic engineering. In theory sheep and goats cannot produce viable offspring when crossed but specimens do turn up which are purported to be geeps. There seems to be little need to cross sheep and goats as there is sufficient genetic variation within each species already.

(b) CROSSING BREEDS

This is the most common technique used throughout the world and is well demonstrated in the British sheep industry where hill sheep breeds (e.g. Scottish Blackface) are mated to rams of specialist 'crossing' breeds (e.g. Bluefaced Leicester) to produce breeding ewes (e.g. Greyface) for the lowland. These lowland ewes are then mated to meat-breed sires. There are many variations of this basic system.

Crossbreds generally perform better than the basic purebreds in reproductive traits. This was well demonstrated by data from 34 800 litters of pigs from 800 farms in Britain. Here purebred dams carrying crossbred piglets produced 2% more at birth, 5% more at weaning and 10% better total litter weight at weaning. When the dams themselves were crossbreds as well as their litters, they produced 5% more pigs at birth, 8% more at weaning and 11% better total litter weight weaned.

A major practical problem with crossbreeding is remembering at what stage of the cross each animal is. This means that a recording scheme is essential for parentage as well as performance. The sophistication of the scheme can vary with the aims and the needs of the breeder, but it is important to stress that the programme should be planned.

(c) CROSSING STRAINS OR LINES
This is where strains, lines or families are crossed within or between populations.

(d) CROSSING INBRED LINES
Here specially produced inbred lines are crossed within a population. It is sometimes called in-crossbreeding.

WHAT TO DO AFTER THE FIRST CROSS (F_1)
Breeders are often concerned about what to do after the first (F_1) cross. To which animals should the F_1 females be mated, to carry on the programme? There are a number of possibilities, and in discussing these, letters are used to denote the breeds or strains involved. Here the word 'breed' is used in a general sense to cover breeds, strains, lines and so on. The basic first cross is A× B to give C:

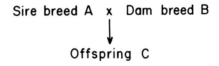

```
Sire breed A  x  Dam breed B
                 |
                 v
          Offspring  C
```

ALTERNATIVE 1

```
   A  x  B
   |
   v
   C  x  (Sire  breed  A or  breed  B)
```

Here C is crossed with a sire from either A or B. This is called backcrossing, in which the breeder goes back to the original parent breeds. Inbreeding would occur if the actual parent were used again, so usually another sire of the same parent breed is chosen instead. In some animals, particularly sheep, C would be called a halfbred as in the British example:

Border Leicester×Cheviot = Scottish Halfbred.

An Australian example is:

English Leicester×Merino = Halfbred.

The backcross of the halfbred to the Leicester would give a three-quarter-bred (i.e. the proportion of the longwool breed dictates the cross). The backcross to the Merino (i.e. quarter longwool) would be called a 'come-back'. There are many local names for different crosses, depending on the country.

ALTERNATIVE 2

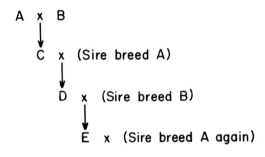

This is an extension of Alternative 1 where A and B are used alternately. This is called crisscrossing and can proceed continuously.

ALTERNATIVE 3

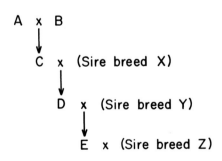

Here a whole range of different sire breeds are used in a planned rotation. This is called 'rotational crossbreeding' and obviously can be quite complex. It is probably most commonly used in pig breeding. Such techniques are very useful in forming large reservoirs of genetic variation on which selection programmes can be imposed.

ALTERNATIVE 4

*Inter*breeding (not inbreeding) the F_1 crossbred C. Here the population is closed and selection of male and female parents is made within it. A modification of this technique is to carry on interbreeding the females, ignoring the stage of the cross but using F_1 sires all the time. This has been used in sheep breeds as, for example, in the New Zealand Halfbred which is an officially recognised breed (an interbred English Leicester× Merino). For greatest success, interbreeding must be accompanied by positive selection. There is substantial evidence to show that, without selection, the superior performance usually found in F_1 over the parent generation is reduced to 50% in the F_2. Selection can be used to counteract some of the decline from the F_1 to F_2 and selection of the F_2 can stabilise the cross. There are many examples of breeds that have been developed by this technique. Some examples are:

* Colbred sheep (Britain) – from crosses between East Friesland, Clun Forest, Border-Leicester and Dorset Horn.
* Coopworth sheep (New Zealand) – crosses of Border-Leicester and N.Z. Romney.
* Australian Milking Zebu cattle (AMZ) (Australia) – from crosses between Jersey and Zebu.
* Santa gertrudis cattle (USA) – from crosses among Shorthorn and Brahman cattle.
* Luing cattle (Scotland) – from Shorthorn and Highland cattle crossing.
* Brangus cattle (USA) – from Brahman and Angus breeds.
* Minnesota Lines of pigs (USA) – from 12 basic breeds.

OUTCROSSING

A breeder makes an outcross when he brings in some new genetic variation – often called 'new blood' – into his flock or herd, and this is usually done by buying in a new sire. The magnitude of the outcross depends on how drastic a change is needed. A breeder may buy a sire from another breeder with a similar programme – this would be called a mild outcross – or he may obtain a sire from a vastly different source and make a more severe outcross.

Although crossing brings in new variation, there may be an increase in the observed uniformity in the F_1 progeny, especially if the inbreeding level of the population had reached a fairly high level before the outcross. This observed F_1 uniformity may not last as the genetic variation is exploited through selection. Outcrossing may often appear to be a 'crash programme' of improvement depending on how mild or severe the outcross is.

BACKCROSSING

As described earlier, this is where a crossbred offspring is bred back to one of its parents, which are usually purebreds. It is often hoped that backcrossing will hold some of the benefits of the F_1 cross.

TOPCROSSING AND GRADING UP

These two techniques are very similar. A topcross is made when a breeder goes back to the original genetic source of the breed or strain for some new genetic material. An example would be Angus breeders from America or Australia returning to Perth (Scotland) to buy a stud sire. These breeders would return home with a topcross.

Grading up is where one breed is changed (graded up) to another by continued crossing. It has been widely used throughout the world where 'native' stock were graded up by a number of crosses with registered sires of improved breeds. Most breed associations accept four generations of crossing with a registered sire as purebred status.

A grading-up programme would work like this:

Unspecified original Female   x   Registered Sire

↓

(50% pure) Female   x   Registered Sire (F_1)

↓

(75% pure) Female   x   Registered Sire (F_2)

↓

(87.5% pure) Female   x   Registered Sire (F_3)

↓

(93.75% pure) Female (F_4)

The F_4 generation female that is 93.75% pure is acceptable as a purebred. This process can take many years as it relies on getting female offspring from each generation on the way through to purebred status. There are usually no performance specifications laid down in grading-up programmes. Obviously, the greatest success will be achieved by using top proven sires for both pedigree and performance.

MATING LIKES

Mating likes is also called 'assortative mating'. It is a very old technique and is still used today. In theory it means more than mating best to best; it must also mean mating worst to worst and average to average. These terms 'best', 'average' and 'worst' do not always refer to productive traits: they are generally more applicable to the phenotype of the animal. It is a technique usually confined to mating best to best and is generally concerned with visual characteristics.

Lush[9] warned against confusing assortative mating with inbreeding. The former is mating animals that have similar looks, while the latter is mating animals that have similar genes. He also pointed out that mating likes was not efficient in altering gene frequency compared to other selection and mating methods.

MATING UNLIKES

Mating unlikes is sometimes called 'negative assortative mating' or more commonly 'compensatory mating'. Here the deficiencies in the characteristics of one animal are balanced by the superior characteristics of another animal. It is a common correction technique and again refers usually to physical traits. An example would be the mating of the historically famous sheepdog 'Roy' (described as having free eye) to 'Meg' (with strong eye) to produce 'Old Hemp' in 1893. This mating by Adam Telfer was the foundation of many Border Collie strains found throughout the world today. Old Hemp was described as 'balanced' by dog men.

Mating unlikes aims to 'even up' the population and exploits a lot of regression to the mean from the extremes of the traits concerned. As with mating likes, any progress made is soon lost because, after the techniques are stopped, the population quickly returns to its original state of variability under random mating.

Hybrids and heterosis

In the previous sections the terms hybrid, heterosis and hybrid vigour have deliberately not been used. This was because of the need to define them carefully to avoid the confusion found regularly in practice. The term 'hybrid' is greatly abused and extra confusion came in when commercial cornbreeding companies in the USA went on to breed poultry. The word hybrid became part of commercial trade-names such as either 'hibrid' or 'hybred'.

Chambers' Dictionary of Science and Technology defines a hybrid as 'the offspring of a union between two different races, species or genera; a heterozygote'. But even this is not clear enough when referring to farm animals. The basic theory is based generally on the facts that when animals are crossed, provided that they differ genetically (as in races, species or genera) then a phenomenon called *heterosis* occurs. This is called positive heterosis if the offspring are better than the mean of both parents, and negative heterosis if the offspring are worse than the mean of both parents.

It is important to remember that the genetic differences between parents should be wide to get heterosis and that either positive or negative heterosis can be expressed in the cross. Not all crosses show positive heterosis but it is this positive heterosis that is called *hybrid vigour* and is of special interest to animal breeders. In animals an offspring rarely is better than *both* parents, or an improvement on the superior parent. In farm livestock a hybrid or an animal showing hybrid vigour is one that is better than *the mean* of both parents (the mid-parent mean). This is summarised in a diagram in fig. 28.

ANIMALS

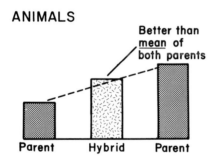

Fig. 28 **Hybrid vigour in animals**

It is important to stress that in measuring heterosis all the animals (i.e. both the parents and the offspring) should be compared in the same environment. Often crossbreds are produced to perform under different conditions to the parents, and in this case it should not be inferred that the crossbreds exhibit hybrid vigour unless they were properly compared with the parents in a common environment. So often, crosses between breeds do not show hybrid vigour as defined above; their performance often equals the mid-parent mean. Hybrid vigour is usually at its maximum in the F_1 and then is halved in each subsequent backcross to either parent.

The breeder's term of 'nicking' can also complicate the discussion of heterosis when it is used to denote hybrid vigour. It is often used in crosses between families or breeds where no real hybrid vigour occurs: it may be used by breeders to describe a successful mating. Nicking, however, is an important phenomenon and is covered under 'combining ability'.

HYBRID VIGOUR IN FARM LIVESTOCK

Some general estimates of how much extra production can come from hybrid vigour are shown in table 13.

Table 13   **Some general estimates of hybrid vigour for traits in farm animals (expressed as a percentage)**

Dairy cattle	Milk yield	2–10
	Fat yield	3–15
	Solids-not-fat	0
	Feed conversion efficiency	3–8
	Birth wt	3–6
	No. live calves per calving	2
Beef cattle	Calving rate	7–16
	Viability of calves	3–10
	Calves weaned per cow mated	10–25
	Birth wt	2–10
	Weaning wt	5–15
	Post-weaning gain	4–10
	Feed efficiency	0–6
	Carcass traits	0–5
	Age at puberty (hybrid younger)	5–15
Sheep	Barrenness (hybrids less barren)	18
	Lambs born per ewe lambing	19–20
	Lambs weaned per ewe joined	60
	Lamb survival	10–15
	Birth wt per lamb	6
	Growth to weaning	5–7
	18-month wt	10–12
	Carcass wt	10
	Fleece wt	10
Pigs	No. born per farrowing	2–5
	No. weaned	5–8
	Total litter wt weaned	10–12
	Growth	10
	Carcass traits	0–5

COMBINING ABILITY – GENERAL AND SPECIFIC

This subject mainly applies to poultry breeding and has not been widely used in large farm animals. It really covers some further methods of selection for conditions where non-additive genetic variation is more important than the additive portion. This is where the D and I components are more important than the G component in the equation on page 52. Here dominance and epistasis are exploited to produce commercial hybrids.

Lerner[20] classified selection into:

(a) Intra-population selection i.e. within a population.
(b) Inter-population selection i.e. between populations.

This is shown in fig. 29 where a population is made up of separate strains or lines, often referred to as genetic isolates. There are three isolates called strains A, B and C that are present at the start of the programme (generation 0). On the basis of a test for appropriate traits, A and B are kept and C is discarded. Then at generation 1, there are A, B and the newly-introduced strain D. On the basis of a test A is discarded so that B and D go on to the next generation and join E, the newly introduced strain. In generation 2, B is discarded and so on.

The main point is to determine how each strain is selected. This can be done by progeny testing within each strain where the best sires are used and the worst culled. This then keeps the strain going. Generation intervals are long because the progeny from the progeny test are not used for breeding. The next step is to cross the strains in all possible combinations (and their reciprocals) to see which ones could be marketed as useful hybrids.

These programmes can become very complex and are used where the lines are fairly inbred. One important method is called 'Recurrent selection' (or RS) where selection is carried out in one line on the basis of its cross performance with another line (usually inbred). This inbred line is regarded as a constant. 'Reciprocal Recurrent Selection' (or RRS) is another method based on intra-population selection using test-cross performance lines.

Thus in all these programmes the breeder may find a strain, a family, or even an individual that gives good results when crossed with *all* the others. This is a *general* combining ability. Alternatively, a strain, line or individual may only produce good results with certain others: this is *specific* combining ability. Both these types of combining ability are used to produce commercial hybrids.

It is easy to accept that hybrid poultry are not worth breeding from as their merits are not guaranteed to be passed on. A commercial hybrid is a 'disposable' product and has an air of planned obsolescence associated with it. It should be used and then thrown away for a better model the following year. This attitude of commercial poultry breeders ensures that breeders face a continuing challenge.

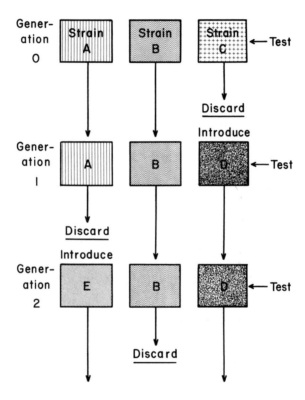

Fig. 29   **Selection in a population made up of different genetic isolates**

IV Breeding in Practice

Practical breeding plans

This section is about 'what to do' in a breeding programme. It can only deal with the general principles as what is done depends entirely on the special needs of each situation. It often seems in practice that each problem is a special case and discussing general solutions may seem of limited value. However, this discussion illustrates the general plan of attack and it is included for this reason. Specific texts (listed in each section) should be consulted for fuller discussions of each class of livestock.

The main principle behind an improvement strategy for a flock or herd is to appreciate that there are four pathways along which the attack can be made. The breeder can concentrate selection on:

(a) Males to breed males.
(b) Males to breed females.
(c) Females to breed males.
(d) Females to breed females.

As discussed earlier in progeny testing and AI, the male is usually responsible for more offspring than any one female so pathway (a) is a major one for improvement. Likewise, pathway (c) is important and is seen in the contract mating of high-merit cows to breed dairy sires for testing. Pathways (b) and (c) together would be useful again in contract mating, where for example in dairy cattle the daughters of the best proven sire in turn became bull mothers. Pathway (d) is perhaps the least powerful unless widespread use of embryo transfer is used.

IDENTIFICATION
The success of a breeding programme depends so much on the accuracy of the data collected and here the system of identification can often be

the limiting factor. Lost or misread tags mean lost data and wasted effort, so identification systems must be of a high standard and, where possible, duplicated for safety.

The following are some methods currently used to identify farm animals:

DAIRY AND BEEF CATTLE

Permanent identification
* Metal (aluminium or brass) ear tags – readable at very close quarters under restraint.
* Plastic ear tags – for individual or group identification and readable at about 3–5 m. Wide range of sizes and colours available.
* Fire brand on hide – readable at 3–5 m if carefully done and the hair is clipped off. Used more in beef than dairy cattle.
* Caustic-burnt hide brand – readable at 3–5 m if carefully done. Used more for dairy than beef cows.
* Freeze brands – readable at 3–5 m if carefully done. Usually require a skilled operator to apply for best results. Brands only show up on dark-coloured cattle as the branded hair grows white.
* Plastic tail, hock and ankle tags – readable close up. Used for dairy cows where numbers are to be read in the milking shed or bail.
* Ear tattoos – readable close up only with animal under restraint and in white ears. Great risk of becoming illegible.

Semi-permanent identification
* Neck tags of various types – metal, plastic, etc., suspended on chains or nylon cords. Readable from 3–5 m.
* Hair dyes or bleaches to write large numbers on the side of the animal – readable from 3–100 m.

Temporary identification
* Paint or spray marks or numbers – readable from 3–100 m.
* Stick-on labels, self-adhesive.

SHEEP

Permanent identification
* Metal (aluminium or brass) ear tags – readable very close up under restraint.
* Plastic ear tags – readable at 3–5 m for individual or group identification. Wide range of sizes and colours available.

* Ear notches – readable at 3–5 m. An example of a system for numbering using notches is shown in fig. 30.
* Ear tattoos – readable close up under restraint in white ears only. Great risk of becoming illegible.
* Fire brands burnt on to the horns of horned breeds – readable close up under restraint.

Semi-permanent
* Neck tags made from plastic, wood, hardboard or metal and attached by cord. Re-usable. Readable from 3–5 m.
* Plastic-covered coloured wire 'twister', inserted through a hole in the ear or through the metal tag. Many different colour combinations can be twisted together.
* Paint brands for individual or group identification. Use very sparingly and only approved scourable (washable) paints and raddles should be used.

Temporary identification
* Paint, crayon or spray brands – again used sparingly and approved as scourable.
* Tie-on tapes, labels, ribbons, clip-on coloured clothes pegs, etc.
* Chalk raddle marks – used sparingly and approved as scourable.
* Stick-on labels.

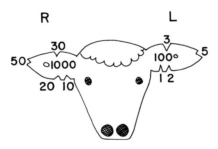

Fig. 30   **A system of ear-notching for sheep**

PIGS

Permanent identification
* Metal ear tags – readable very close up under restraint.
* Plastic ear tag – readable at 3–5 m.
* Ear-notching systems – readable close up and up to 3 m. An example of a numbering system is shown in fig. 31.

* Ear tattoos – readable close up under restraint in white ears only.
* Skin tattoos – used prior to slaughter to identify the carcass.

Temporary identification
* Paint or spray marks – readable at 3–5 m.

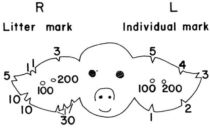

R   L

Litter mark   Individual mark

Fig. 31 **A system of ear-notching for pigs**

POULTRY

Permanent identification
* Wing tags or wing bands – metal or plastic for individual or group identification, readable close up or at 3–5 m if large tags are used.

Temporary identification
* Paint or spray marks – readable at 3–5 m.

DAIRY CATTLE BREEDING

In dairy cattle the main records are associated with milk yield, regular calving and milk quality. Most countries have official milk recording services and national herd-improvement programmes. These should be consulted for details. Milk quality assessments require herd testing and laboratory services.

Milk yield has a low to medium heritability indicating that selection gains will be achievable although perhaps slowly. It also has low to medium repeatability indicating that individual selection will make some progress. This means that the cows with proven performance in the herd must be kept to breed replacements. The best of these cows should then be used as the dams of young bulls to be progeny tested. Fully-proven sires should be used extensively throughout the herd. Some top cows may be offered in sire proving schemes to be mated to these young bulls with very high potential.

Milk quality traits are generally highly heritable and can be improved directly by selection. The genetic correlations among quality components are high and positive. The only problem relationship is the negative genetic correlation between yield and fat percentage – as yield goes up, fat percentage goes down.

Aim 1: To identify and keep the best dams

Action
* At calving, record:
    Calf no. and year born (permanent tag)
    Dam no.
    Day born
    Birth weight (optional)
    Sex of calf
    Calving difficulty (using a score such as
    1 = not seen; 2 = seen but no assistance;
    3 = slight assistance by one person with no aids;
    4 = considerable assistance using mechanical and veterinary help).
* Record milk yield by volume or weight. Milk fat (MF), solids-not-fat (SNF) and protein are the main quality constituents to consider. Specify the number of times milked per day as this greatly affects total lactational yield.
* Record the dairy temperament of each cow. The aim is to cull those of nervous disposition that cannot adapt to the system, slow milkers, and those prone to mastitis, etc. These animals will generally also be low producers but a double check (on yield and temperament separately) is worthwhile.
* Record all heat periods of each individual cow after calving and when she is served. The calving date can then be predicted and the calving interval calculated.

The aim is to build up lifetime records on each cow and cull the low yielders as well as any that are difficult to handle or prone to disease within the particular farming routine used. Further culling can be done for milk quality (mainly fat) if this is economically important.

Aim 2: To select the female replacements for sexual maturity pregnancy and ease of handling

Action
* Cull all heifers that do not show oestrus or become pregnant after a restricted mating period. The length of the joining period with the bull (natural service or AI) depends upon how important the calving spread is, e.g. is it a short calving spread or is it calving all the year round? Once the herd is in milk, again cull on ease of handling with special emphasis on the udder, teats and speed of milking.
* Only the calves with the best genetic background should be reared.

* Mate these 'best-bet' heifers to the best proven bulls that are available. Often heifers may be mated to beef sires for their first calf to reduce the risk of calving trouble. This will affect genetic gain by lengthening the generation interval but has to be balanced against the risk of injury or death of the heifer.

Genetic correlations are around zero between growth rate, skeletal size and milk yield so little would be gained by selecting for size. Larger animals would in any case have greater maintenance costs although their beef salvage value may be greater. In practice farmers would usually be more keen to cull the poorly-grown heifers for management reasons knowing that they would not make good cows – regardless of their genetic potential.

Aim 3: Use only the top proven (progeny tested) sires available

Action
* Progeny testing of sires is an essential part of dairy cattle breeding because dairy traits are not highly heritable and are expressed only in the female. The fact that progeny testing lengthens the generation interval has to be accepted. In large breeding programmes, however, such as national breeding schemes, the very high selection intensities that can be achieved in bull-proving schemes can help to counteract the effects of longer generation intervals. Generally, the small dairy herd must rely on AI and leave the responsibility for proving bulls to larger organisations. Each dairying country has its own system of evaluating dairy bulls but they all use the concept of providing a form of Breeding Value for the sire – in other words a prediction of his ability to produce superior daughters. An example described in some detail below is the Improved Contemporary Comparison (ICC) used in England, Wales and Scotland.

THE IMPROVED CONTEMPORARY COMPARISON (ICC)
This is the comparison of a bull's daughters with the daughters by other bulls (i.e. contemporaries) calving in the same herd at the same time and treated similarly. This approach of using a contemporary comparison was developed in the 1950s to eliminate the effects of environmental differences between herds and was modified into the Improved Contemporary Comparison (ICC). Here, the calculation takes into account and adjusts for the age of the contemporaries, their genetic merit and the season of calving. This means that results from

Table 14 Example of calculation of Improved Contemporary Comparison (ICC) for a Limited Usage (LU) sire. Milk yield in kg.

Herd-Year-Season batch of records (1)	Daughters of the Limited Useage sire		Daughters of other sires (contemporaries)					Differ-ence (3) – (8) (9)	Effective number of daughters (weighting) (10)	Weighted differ-ence (11)
	Number of daugh-ters (2)	Age and month correc-ted milk yield (kg) (3)	Number of contem-poraries grouped by sire (4)	Average age and month correc-ted yield (kg) (5)	ICC of sire of contem-poraries (or G for a LU bull) (kg) (6)	Correc-ted yield of contem-poraries (7)	Average for the herd/season (8)			
A	10	4500	4	4250	+250	4000	4000.0	+500	3.75	1875
			2	3800	−200	4000				
B	5	5500	10	4800	−200	5000	4980.4	+519.6	4.17	2167
			10	5001	0	5001				
			5	5300	+400	4900				
									7.92	4042

different herds in different years can be combined into a single figure as seen in the calculation in table 14.

* Column 1 shows two herds (A and B) milked in the same year and the same season, described as two herd-year-seasons. The data are handled in the same batch of records.
* Column 2 shows the number of daughters of the sire being tested (called a Limited Use or LU sire) in each herd-year-season group of daughters.
* Column 3 shows the average milk yields for the LU sire's daughters, corrected for age and month of calving.
* Column 4 shows the number of daughters by other sires milked in each herd A and B in the same year and season.
* Column 5 is the average milk yield of the contemporaries, again corrected for age and month of calving.
* Column 6 is the known ICC of the sire of the contemporaries and is either added (if minus) or subtracted (if plus) from column 5. This gives the values in column 7.
* Column 8 is thus the average i.e. a weighted average corrected yield of the contemporaries. This average is then subtracted from the average of the daughters of the LU bull to give the answer in column 9.
* The next step is to carry out an adjustment for the different number of daughters and contemporaries involved. The more daughters, the more reliable is the mean and hence the more reliance can be given to the final ICC. This is done by the formula:

$$\frac{\text{No. of daughters} \times \text{No. of contemporaries}}{\text{No. of daughters} + \text{No. of contemporaries}}$$

$$A = \frac{10 \times 6}{10 + 6} = \frac{60}{16} = 3.75$$

$$B = \frac{25 \times 5}{25 + 5} = \frac{125}{30} = 4.17$$

These are shown in column 10.

* Columns 9 and 10 are multiplied together to give the answer in column 11 which is the weighted difference.
* Therefore the 'apparent merit' of the sire is:

$$4042 \div 7.92 = 510.4$$

Thus the LU bull on test here appears to be improving yields by 510 kg over contemporary daughters by other bulls. Now the question remains as to his genetic worth expressed by the ICC. To get this, two further refinements are carried out. The first is to scale the comparison to a base year – a concept of comparing current merit to a base of zero genetic value. The second is to use a weighting (a regression factor) for the number of effective daughters in the comparison: the more there are the greater is the accuracy.

The final ICC of the bull used in this example after making these adjustments is +209 kg of milk for 7.92 effective daughters. This means that, on average, daughters of this bull will exceed the base value by 209 kg of milk. As more effective daughters come into milk, the bull's ICC or proof will alter.

The ICC is described as a 'transmitting ability'. This is half a Breeding Value (BV).

CROSSBREEDING
Crossbreeding has not been widely used to improve dairy cattle traits in developed countries where emphasis has been mainly on within-breed selection. An important exception, however, is the widespread use of Jersey × Friesian cows in New Zealand where the crossbreds are bred back to Jersey then Friesian sires in rotation. In developing countries crossbreeding has great potential to improve yield and dairy temperament to native cattle that retain heat and insect tolerance.

For further reading on dairy cattle breeding see references [21], [22] and [23].

BEEF CATTLE BREEDING
In beef cattle the main records needed are pedigree (parentage), growth and calving data. Carcass data collected after slaughter may also be obtained. Most countries have official beef recording schemes and these should be consulted for details of requirements. Examples are the Meat & Livestock Commission (MLC) in England and Wales; Performance Registry International in USA; Record of Production (ROP) in Canada; Beefplan in New Zealand: National Beef Recording Service in Australia.

Aim 1: To identify and retain the best dams

Action
* At birth record:
     Calf number (permanent tag)
     Dam number

Day born
Birth weight (optional)
Sex of calf
Calving difficulty (use a score as in dairy cattle).
* At weaning record:
  Weaning weight (actual)
  Day weaned – to calculate age at weaning from day born.
* Adjust the actual weaning weights for the main environmental variables, e.g. age of calf, sex of calf and age of dam. This then produces a corrected or adjusted weaning weight and can be used as the main trait for selection.

In many breeding schemes, the weights at weaning (whenever the calves are weaned within a limited spread), are converted into a standard 200-day weight which is corrected for environmental effects. The usual technique here is to take 200 days' worth of the gain between birth and weaning. For this, the birth weight is needed or else some accepted standards are taken. The formula used is:

$$200 \text{ day weight} = \left(\frac{\text{weaning wt} - \text{birth wt}}{\text{age in days at weighing}} \times 200 \right) + \text{birth wt.}$$

This 200-day weight can then be corrected for age of dam by adding 15% extra for a 2-year-old dam, 10% for a 3-, 5% for a 4- and nil for a 5-year-old or older dam. Normally the male and female calves are listed separately so sex corrections are not used. However, a correction for sex (i.e. for the lower mean weight of females below males) could be made.
* Select for 'weight of calf weaned' as a character (WCW) as probably the best all-round measure of production from the cow-calf enterprise. It includes both fertility and growth rate. This trait can then be used to rank each dam by comparing each individual with the average of the group each year, (i.e. the deviations).
* The WCW deviations from average can then be built up into an index, after an adjustment is made for the number of records of each dam so that they are all compared equally. Then all the animals can be ranked on the basis of the one index value. The names for these indexes vary depending on the different schemes in each country. Thus in the USA they are called 'Most Probable Producing Abilities' (MPPA), and in other countries they are a 'Lifetime Productivity Index' (LPI) or 'Weaning Index' (WI).
  WCW as a trait has medium heritability, it is easy to measure and it

responds to selection. It is, however, recognised as a complex trait as it covers the dam's reproductive efficiency, freedom from birth problems, lactation, survival, growth and mothering. Nevertheless it is the main money-making trait in beef cattle.

Aim 2: To select for post-weaning growth in replacement heifers and bulls

Action

* It is usually easier to select for a weight rather than a gain, simply because gain is a calculation based on two weights. The two things (weight and gain) are basically the same in any case. Unfortunately, under some feeding systems such as at pasture, weight and gain have a different heritability so different strategies are needed.

    The post-weaning gain or weight is simply added on to the already-corrected weaning or 200-day weight.

* The most common approach is to select for weight at 400 and 550 days of age as further stages along the growth curve. This can be done most effectively in a *within-herd* performance test where all animals are treated alike and the heaviest ones are kept. As 200-, 400- and 550-day weights are genetically correlated, they will all be improved as will birth weight. Selection for gain would similarly affect these characters.

* Select for structural soundness and breed acceptability at the end of the performance test using scoring systems, after the ranking on performance is known. A combined weight and physical inspection score can perhaps be used for an overall-merit assessment.

* If a choice had to be made between 400- and 500-day weights, the 400-day weight would be better as it is nearest to puberty and decisions on which were the best animals could be made before mating as yearlings. If they were not mated until 2-year-olds, then the later 550-day weight would suffice. Selection at the younger age would help to shorten generation interval.

Aim 3: To select for reproduction in the heifer replacements

Action

* Identify oestrous activity in the heifers by using a harnessed vasectomised (teaser) bull if they are not to be mated as yearlings. If they are to be mated, then those that conceive early may be identified by a harnessed entire bull or by pregnancy diagnosis by a veterinarian.

* Decide whether the heifers are heavy enough to join with a bull and what is an optimum weight for mating. This depends on the farming system, breed, growth potential up to calving, etc. Many breeders take the view that the bull can decide. Here all the heifers may be joined with the bull and those that are too light, or not pregnant or both can be culled. The culling point can be set by the number of replacements required.

* The main action is to identify good growth rate, early oestrus and regular calving. There may be some concern over what is meant by 'good' growth because excessive rates of growth leading to over-fatness may be harmful to subsequent maternal performance. Growth is highly heritable but oestrous activity and pregnancy rate are not. However, weight and reproduction are related phenotypically so putting emphasis on growth will ensure the expression of reproductive potential. Putting selection pressure on the development of puberty is a very sound aim to ensure the improvement of overall reproductive performance – although progress may be slow.

Aim 4: To select for carcass weight and reduce fat

Action
* The cold carcass weight can be obtained from the abattoir or processor and is the major component of profit. The other useful and easily obtained measure is fat depth at a defined spot e.g. the 13th rib. As these traits are only available after slaughter, progeny testing could be used to identify superior sires and these could then be used widely through AI within a herd. It would be doubtful if the cost and the resultant longer generation interval would be counter-balanced by the genetic gain. Large numbers of sires would have to be compared e.g. at least 5–10, and sufficient progeny of one sex (10–15) per sire would have to be examined.

* A possible policy could be to select bulls within terminal sire breeds (e.g. Charolais and Limousin) for carcass traits by examining the bulls themselves by electronic scanning of eye muscle, followed by progeny testing of their sons (as steers) for carcass traits and their daughters for calving difficulty. This would be a large programme requiring many progeny, 200–300 per sire.

* Assessment of fat cover on the live animal by the use of ultrasonics would be worthwhile, especially on sires. This would at least indicate what variation (phenotypic) was present. This assessment could be used at the end of the within-herd performance test.

CROSSBREEDING

Crossbreeding can be used widely in beef production especially in commercial cow/calf operations where the dam would be an F_1 cross and a large terminal-sire breed would be used to produce the slaughter generation. Crossbreeding may be exploited in more complex schemes (described on page 96) to strive to maintain hybrid vigour in maternal and calf traits.

For further reading on beef cattle breeding, see references [22, 24] and [25].

DUAL-PURPOSE CATTLE BREEDING

Aim: To select for both milk and beef traits

Action
* Select for milk production and quality characters as described for dairy cattle.
* Select female replacements for 400-day weight from a within-herd performance test and then join these with the bull. Then those that became pregnant and calved without difficulty could enter the milking herd.

  The main concern is whether there is an antagonism between meat and milk characteristics. Generally there has been shown to be no significant genetic correlation between meat and milk traits so the two aspects have to be considered separately in an improvement programme.
* Select males for growth in a *within-herd* performance test. This would identify the best-grown sires, and those from dams with good lifetime dairy records could be progeny tested before widespread use. If the male calves had to be castrated, then they could be used to assess growth and carcass traits of their sires.

For further reading see references [21, 22.]

SHEEP BREEDING

In sheep breeding, the most important records required concern pedigree and fertility, growth, wool production and also, perhaps, aspects of wool quality. Many countries have official national recording schemes and these should be consulted for details. Examples are the Meat & Livestock Commission's sheep recording service in Britain and Sheepplan in New Zealand. See Owen[26] for a review of world schemes.

Further comment is necessary on the records needed and used to describe flock performance, especially reproductive traits. The basic statistics described here are those of Turner and Young[17] and are fairly universal. The only exception is the term *joined* which is used to define ewes put with a ram in the same field. This is different to ewes *mated* which are those actually mated or served by the ram. The basic statistics needed are these:

> No. ewes joined with ram (EJ)
> No. ewes lambing (EL)
> ∴ $(1 - EL/EJ) \times 100 = \%$ barrenness

> No. ewes lambing multiples (ELM)
> No. lambs born (NLB)
> ∴ ELM/EL = multiple birth rate

> No. lambs weaned (LW)
> ∴ $(LW/NLB) \times 100$ = lamb survival
> or $100 - (\%$ lamb survival) = lamb mortality.

The most confusing statistic used by sheep breeders is 'lambing percentage'. By this they can mean any of the following:

> LB/EL = Lambs born/ewes lambing
> Live LB/EL = Live lambs born/ewes lambing
> LB/EJ = Lambs born/ewes joined
> LD/EL = Lambs docked/ewes lambing
> LD/EJ = Lambs docked/ewes joined
> LW/EJ = Lambs weaned/ewes joined.

DUAL-PURPOSE SHEEP BREEDING

Aim 1: To identify and select the best ewes

Action
* At birth record:
> Lamb number and year born (use permanent tag)
> Dam number
> Lamb sex
> Day born
> Birth rank (single or multiple)
> Birth weight (optional)

* At weaning record weaning weight (actual) and day of weaning.
* Select for weight of lamb weaned (WLW) which includes all aspects of reproduction, maternal ability and lamb growth rate. It will be necessary to correct WLW for environmental variables such as:
  Age of lamb (correct to mean weaning age of group)
  Birth/rearing rank (correct all multiples to single basis)
  Sex of lamb (correct to male basis)
  Age of dam (correct to mature-dam basis)
* Use WLW to evaluate the productive ability of the dams and their annual performance can be built into a productivity index.

WLW is a characteristic with a medium heritability and is economically very important. All the fertility and survival traits have very low heritability and progeny testing is the only way to identify superior sires. Here the disadvantage of a longer generation interval may have to be accepted as the search for superior rams becomes necessary. In the meantime, selection on the female side can carry on by only keeping replacement rams (and ewes when possible) out of dams that have a high performance expressed as WLW, or NLB, if this suits the particular environmental circumstances.

Aim 2: To select for growth (live weight)

Action
* Record these most important weights:
  Weaning weight (4 months old)
  Yearling weight (14 months old) or hogget 18-month weight (1st joining) 2-tooth or shearling.
  These traits are medium to highly heritable and in the female they are associated (both genetically and phenotypically) with oestrous activity and fertility, especially at the yearling stage.
* Select on yearling weight and then among the heaviest yearlings further select those that showed oestrus – either to a harnessed teaser ram or an entire.
* Multiple-reared yearlings may still express an environmental handicap so it may be necessary to make selection decisions *within* multiple-born and single-born groups and correct them for birth-rearing rank.

This selection for growth is carried out by a within-flock performance test, separately for rams and ewes. The decision has to be made about which stage of growth (i.e. at which weight) decisions should be made, remembering that all growth points are related. There is good evidence that 6-month or 12-month weight is more heritable than weaning weight (2–3 months); therefore weaning weight as well as post-weaning gain would be improved by selection for some later weight, although waiting for the older-age data would take longer.

The animals selected on growth can then be culled for physical defects such as teeth, jaws, feet, reproductive organs and general health. Progeny testing for growth traits need not be considered as they are highly heritable and it would lengthen the generation interval. However, using ram lambs instead of 18-month (2-tooth) rams would reduce the generation interval. These ram lambs would have to be used before their yearling fleece weight was known so no selection could be done for wool production. Fleece traits are highly heritable and respond to direct selection and could be given attention later on in the programme. This would depend greatly on the relative importance of meat and wool.

Aim 3: To select for fleece weight and quality

Action
* Select for fleece weight at the yearling stage (their first fleece). This trait is highly heritable and is also highly repeatable, i.e. it is a good indicator of subsequent annual wool production.
* Identify the best animals through a within-flock performance test of both males and females run separately. As live weight and fleece weight are related (genetically and phenotypically) then the two traits can be selected together. The usual way is to cull on live weight, and then cull on fleece weight within the live-weight-selected animals. As fertility and live weight are also lowly correlated (phenotypically and genetically), fertility would also benefit somewhat from the live-weight selection if it were acceptable to have larger sheep with associated effects on stocking rate, etc.
* Select for quality aspects of the fleece at the yearling shearing. Greater attention could be given to the desired quality traits in the ram than in the ewe replacements. Greasy fleece weight should remain the main concern except in specialist wool breeds like the Merino where attention must be given to *clean* fleece weight and other traits.[17]

MEAT SHEEP
Aim 1: To select for growth

Action

* Take a similar approach as for dual-purpose breeds except that more emphasis needs to be placed on growth before weaning but more especially immediately after weaning at the 6- and 12-month stage. In meat sheep, weaning weight is not considered a trait of the dam. Growth traits are highly heritable and progress will be made by within-flock performance testing of ram and ewe replacements.

Aim 2: Select for carcass traits.

Action

* Select for cold carcass weight as the main character. If there is interest in selecting for other traits of carcass composition (provided that they can be obtained), it is necessary to progeny test sires. This would greatly increase the generation interval and slow up progress compared to, say, using the fastest growing ram lambs on the flock, especially on the young ewe replacements. Perhaps a combined performance then a progeny test of the best bets would be worth considering.

    Fatness is another important carcass trait, and electronic methods for measuring fat depth on the ram lamb replacements could especially be considered. Rarely is the extra accuracy from progeny tests for carcass traits sufficient to counteract the costs and time involved.

ARTIFICIAL INSEMINATION

Artificial insemination of sheep is still not widely used in some countries as a means of exploiting top rams. However, there is clear evidence now that greater use of top rams by natural service can be achieved by simply altering the mating ratio from one ram to 40 or 50 ewes to one ram to 100 or even 200. This helps to increase the selection differential and therefore genetic progress.

CROSSBREEDING

Crossbreeding is used extensively in sheep breeding where a well-recognised stratification system has been developed in some countries (such as Britain) where different crossbreds are used to suit different farming systems and market requirements.

    For further reading on sheep see Turner and Young[17] and Ryder and Stephenson.[27]

PIG BREEDING

In pigs the breeder is concerned with reproduction, growth and carcass traits. As feed costs make up such a high proportion of total costs, high efficiency in terms of feed to carcass lean meat is vital for the profitability of the pig enterprise.

Aim 1: To identify the best sows

Action
* Record:
     No. pigs born per litter
     No. pigs weaned per litter
     Weaning weight of each piglet
     Age at weaning
     Total weight of litter weaned
* Record:
     Mating records for each sow
     Day weaned litter
     Day served and return to service
* Cull sows on fertility and maternal ability by using the total weight of litter weaned. These reproduction traits are weakly inherited, as is weaning weight. The performance of each sow can be built into a productivity index over a specified time.
     Note that sows may farrow more than once a year.
* Further culling can be done on physical defects or disease.

Aim 2: Select female replacements (gilts) on growth, conformation and sexual maturity

Action
* Select for growth on a within-farm performance test. This generally means picking the replacement gilts out of the bacon pens, the culls going straight to slaughter. This ensures that the animals are selected on a commercially viable feeding and management regime. Daily gain has medium to high heritability and has a high negative genetic correlation with feed efficiency. Hence fast-growing and efficient animals will be identified. The total feed consumed should be recorded.
* At bacon weight the gilts can be examined for structural soundness and any that do not conceive over a restricted mating period can be culled further.

* The benefits of crossbreeding in fertility, maternal and growth characters in the female have been clearly demonstrated and should be seriously considered in a breeding plan.

Aim 3: Identify the best boars for growth and feed efficiency

Action
* Select boars for growth through performance testing because of the reasons described for gilts. The boars with the best potential on paper (i.e. out of the best-performing sows by the best proven sires) can be selected at weaning, and then sent to some central or national performance test centre outside the breeder's herd. Here the feeding and environmental conditions are kept constant but may differ from the breeder's own system. If a sufficiently large within-herd performance test cannot be organised (as in a small herd) then central testing facilities may need to be used.
* Select the best-growing boars out of the bacon pens as for gilts if they can be left entire up to this stage. Remember that if they are fed in small groups or groups of different sizes, there could be bias caused by 'group effects'. Individual penning or individual feeding with group housing should be the aim for technical accuracy. It must be decided whether to feed on a restricted scale based on weight or on an *ad libitum* system. Whatever system is used, the quantity of feed consumed and analysis of feed for energy and protein level should be recorded.
* If all the males cannot be left entire until the end of the test (pork or bacon weight), then decisions will have to be made on pedigree and performance of relatives to screen prospective animals for testing.
* If different breeds are concerned in performance testing, they should be housed separately. The end of the test is usually a fixed live weight when the boars can be inspected for structural soundness and breed characteristics. Backfat-depth measures are highly heritable so the use of ultrasonic data can be valuable in identifying those animals that had high growth and low backfat.

Aim 4: Progeny test boars for carcass traits

Action
* Progeny testing boars for carcass traits is needed because the important ones can only be assessed after slaughter. The exceptions are backfat depth and eye muscle area that can be measured by

ultrasonics. Most of the important carcass traits are highly heritable and can be measured objectively. The aim is to reduce fat – principally backfat. Backfat and length are negatively correlated (both phenotypically and genetically) so increasing length is a commendable aim. Progeny testing requires large facilities as in other animals and lengthens the generation interval. However, top proven boars can be used through AI in small herds with no testing facilities.

CROSSBREEDING

Crossbreeding is used widely in pigs, especially to develop F_1 sows to improve maternal and growth traits through hybrid vigour. Many new breeds have been developed from these crossbreds.

For further reading on pig breeding see references [22] and [23].

POULTRY BREEDING

The main trait to record in poultry breeding programmes is egg production, often expressed as the 'hen-housed average' (HHA). This is the mean production per bird over the number of birds present (i.e. housed) at the beginning of the period. It also includes mortality over the period.

Body weight is important and birds are easily weighed while suspended in an open-ended funnel. It is also necessary to record the feed consumed because it is the major input (approximately 80%) of the poultry enterprise.

In meat birds, carcass weight is the basic trait and if carcass dissection is carried out, the proportion of breast and thigh meat to total carcass meat is valuable information. Dissection, however, is usually very expensive in terms of labour.

EGG PRODUCTION

Aim 1: To increase egg number, egg size and weight

Action
* Select directly for these traits. They are all basic to the profitability of the commercial laying enterprise and are all correlated both phenotypically and genetically. Increasing egg number can result in the production of a greater number of smaller eggs and each egg will be of lighter mean weight unless some counter action is taken.

These traits are moderately heritable, so respond to selection.

However, exploiting non-additive genetic variation through hybrid vigour has been widely used and most laying strains on the market are hybrids (see page 105).

Aim 2: To reduce body weight and improve feed conversion efficiency

Action
* In egg strains, reduce body weight and hence reduce maintenance costs through a lower appetite and possibly also achieve a greater feed conversion efficiency (FCE). However, FCE would have to be selected for directly to ensure progress.
* Use performance testing as an initial screening operation to identify good individuals (males) and this could be followed by progeny testing. Selection techniques as described on page 105 would be used where possible. The individual bird can be very widely exploited for genetic reasons through the large number of eggs one female can lay, and males can be used widely through AI.

Aim 3: To improve livability

Action
* Adopt the simple approach by only concentrating on the main diseases that do not respond quickly and cheaply to husbandry techniques. Performance test for these by keeping the best-performing animals in the disease environment. This is a difficult area as there are now so many diseases to which birds can be exposed.
* Progeny test to check that the disease resistance has been passed on to the commercial market progeny and is expressed in different environments (husbandry systems).

Aim 4: To improve egg quality

Action
* Select directly for the main traits of importance. These are shell strength (important in collection and storage), shell colour (in some countries consumers prefer brown or tinted eggs to white), yolk colour (bright instead of pale yellow), texture of white, freedom from blood and meat spots.
* Identify the superior parents by performance testing and then progeny testing. These parents can then be used in family selection and exploited through crossing and heterosis described on page 105.

MEAT PRODUCTION

Aim 1: To improve carcass weight and conformation, and to reduce fat content

Action
* Adopt the same techniques as discussed for other poultry traits. Carcass weight and conformation have fairly high heritability and respond to selection through performance testing. Fat (which in the fowl lies inside the body cavity) can be accurately assessed and selected against, depending on consumer preference.

In meat birds, egg production is still important as it is an essential part of multiplying the highly selected birds to meet the market's orders. This problem is often attacked, especially in turkeys, by crossing a sire (meat) line with a female (egg laying) line to produce the market hybrid.

For further reading on poultry breeding see reference [20].

BREEDS AND BREED STRUCTURE

The definition of a breed can only be very general. It is a group of animals, within a species, that has a common origin and certain physical characters that are readily distinguishable. Once these physical traits are removed, e.g. by skinning after slaughter, it often becomes difficult to tell breeds apart. Thus the physical features act like a breed label. Isolation by barriers (e.g. mountains and seas), regulations, social differences among their users and fashion have all helped to keep breeds separated. In genetic terms, isolation caused the genotype to drift apart (genetic drift). Genetic differences *within* breeds can be large and in some cases can be as big or greater than those *between* breeds.

The structure of a breed is important as it controls the way in which genetic improvement flows throughout the breed. Breeds are best visualised as a hierarchy, drawn as triangle.

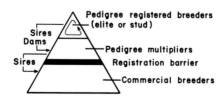

Fig. 32   **Traditional breed structure**

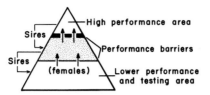

Fig. 33 **An improved breed structure**

In the traditional structure (fig. 32), at the very top of the triangle there are pedigree registered breeders (sometimes referred to as elite or stud breeders). Then there is a layer of other pedigree breeders who multiply the material from the elite breeders.

Below the registration barrier are the non-pedigree commercial breeders who receive the genetic benefits of those breeders above the barrier. Genetic material flows down through this structure, usually by the sale of males. The flow can be hastened by AI so that semen from the elite breeders' animals can go directly into flocks or herds in the base. So the whole system is based on the assumption that the elite breeders are making progress and that this is being constantly released. Usually the elite breeders improve by exchanging males amongst themselves or by importation from outside, e.g. from the home of the breed where there is likely to be another similar structure.

This structure can be criticised from a genetic viewpoint because:

* The flow is one-way and genes cannot flow up into the registered areas from the non-registered part, unless the flock and herd-books are still open. Generally they are closed.
* The barrier is simply a 'registration' barrier and not a 'performance' one. New stud breeders usually have to start by buying surplus registered females (often culls) from other studs. They usually cannot start by using good-performing commercial stock and having them registered.
* The registered flocks and herds are generally made up of small numbers of animals hence the opportunities for selection are greatly restricted. Scope for selection is clearly greatest in the larger flocks and herds in the commercial area although here there are usually practical difficulties in recording large numbers of animals.

A suggested improvement to the breed structure would be that shown in fig. 33 where animals with good performance could flow from base to apex. The former registration barrier is then replaced by a performance barrier.

BREED ASSOCIATIONS

The term breed association is used here to include breed societies, livestock record associations, etc. Lerner and Donald[1] gave an admirable summing-up of the history and role of breed associations where they pointed out that breed associations are often part of the cultural inheritance of many countries. Breed associations are regularly criticised by technical people and this criticism can probably best be summed up in a series of questions such as follows:

> What do breed associations do?
> Are they really needed?
> Why have some of them (e.g. in poultry and pigs) disappeared?

Views tend to be polarised into those 'for' and those 'against'. Here are some examples to illustrate the argument:

Points for
* A breed association takes responsibility for a breed and this is both a physical responsibility (i.e. administration) and a moral one. It is an obvious source of information for performance specifications, sales, standards, exports and imports, etc.
* It is a reliable body for collecting and recording the ancestry of all animals in the breed for all time. It can thus trace the ancestry of any individual back to the source of the breed and hence ensure its 'breed purity'.
* It can provide a focus for breed promotion for members through sales, field days, demonstrations, conferences, advertising, and so on.

Points against
* The effort and expense breed associations spend on recording extended pedigrees and producing flock and herd books is unnecessary. Recording pedigrees without performance is out-dated and serves little purpose.
* Breed associations are usually too concerned with self-preservation. Their councils generally have many more older-established breeders than younger breeders, hence the chances of new ideas and rapid changes are limited.

LIVESTOCK SHOWS

The fortunes and future of livestock shows and breed associations are closely linked and any discussion of the subject among breeders, farmers and scientists again reveals that views are generally polarised.

Polarisation mainly occurs between those who believe that animals should only be compared using performance data, and those who believe that physical appearance is adequate for comparison. There are currently plenty of people who believe that both performance and physical data should be used but the question of how it should be done remains to be answered adequately.

Some discussion points for and against showing are:

Points for
* Shows are the 'shop window' for the breed where 'good' specimens, (i.e. approved by the top judges) can be seen by all who are interested – breeders and buyers alike.
* Young breeders can see the ideal to aim for and the only place to identify this aim is in a competitive show.
* Shows provide a meeting place for breeders and buyers. They can become discussion and education areas among the people involved in the industry.
* Shows provide a valuable way of bridging the ever-increasing gap between urban and rural people throughout the world. Town people can see and touch animals and talk to their breeders and owners: this is becoming very important in an age of increasing urbanisation.
* As a result of the open competition at shows, superior animals are identified and these can then go to influence the breed's future either through use in the top flocks or herds (that also support shows) or through artificial insemination.

Points against
* The definition of 'best' is usually based solely on physical form or type, and this is usually strongly biased by personal fancy and rarely proven fact. Indeed, the commercially superior animal may never be exhibited.
* Shows encourage excessive pampering and gross over-feeding which are completely unrelated to commercial practice. As a consequence show results are generally ignored by commercial farmers.
* Any comparison between animals (even if backed by performance data) cannot be valid because of the confounding influences of the different environments from which they came. Comparing animals at a show is really more a comparison of the stockmen who prepared them for exhibition.
* So few animals from a population are exhibited that it cannot be assumed that they are the best specimens of the breed for future

exploitation. Even when as in some shows, sires and a group of their progeny are exhibited, the non-random selection of progeny invalidates the comparison.

There remain many breeders who believe that showing improves their returns sufficiently to provide support for shows. There is an increased interest in improving the design of livestock shows to make the exhibition of stock more related to commercial needs. Already in some quarters there is a change from competitive showing to more of a demonstration of superior animals. There is clearly a demand for permanent areas to provide demonstration information about animals and many have been built in different countries.

CO-OPERATIVE BREEDING SCHEMES

Interest in co-operative breeding schemes is generated by the basic genetic principle that it is possible to apply more selection pressure in a large population than in a small one. There is no reason why small breeders cannot exploit these benefits through co-operation among themselves. The principles of a group-breeding scheme are very simple and are described in fig. 34. This shows herds of different size from

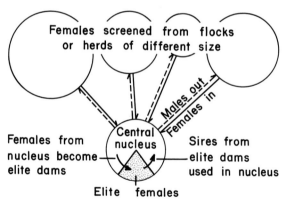

Fig. 34 **Simplified structure of a group breeding scheme**

which females can be screened to form a nucleus and arrows are used to denote the females going into the nucleus and males returning to the contributors.

Further details of these schemes can be described in the following suggested programme for setting up a breeding scheme:

(a) Form a group of interested breeders to discuss the concept and the business, legal and genetic aspects of the scheme – probably in this order of priority.

(b) Each breeder contributes the top performing females in his flock or herd to a central unit (nucleus). This is the concept of 'screening' the population for the good females.

(c) Decide where the nucleus is to be located and how it is going to be managed. The manager is a most important person in controlling the actual level of performance achieved in the nucleus.

(d) Decide on an exchange rate of top females *in* to the nucleus for selected sires *out*. Usually a ratio of four females in: one male out is a useful starting point. This can be based initially on commercial value when genetic merit is unknown.

(e) Continue screening in each contributor's flock or herd and selection in the nucleus. It is vital that this selection is based on productive fact and not fancy. Conformation traits are important as long as they encompass structural soundness, but these traits should be clearly defined for the benefits of all members of the group so that they do not swamp productive traits in the order of priority.

(f) Replacements can be obtained from those bred within the nucleus and from animals screened in. Initially half can be nucleus-bred and half can be screened until the programme develops.

(g) Within the nucleus, the top performing females will acquire elite status as more performance data accumulate. These females must then be mated to the top sires within the scheme to breed sires for use within the nucleus.

It is important to realise that most genetic progress comes from the high selection pressure made possible in the initial screening operation. After that, genetic progress will depend on effective selection, as in any other flock or herd.

An interesting outcome of these group-breeding schemes has been the great educational and extension potential that they have. At group meetings and field days, especially on the annual occasion when all members are present to select their sires, unrestricted argument and discussion can take place among breeders with a common overall interest – to breed better stock. This has certainly been a highlight of the many schemes in both cattle and sheep operating in Australasia.

The possible use of such schemes in developing countries is also worthy of study, because the limited technical expertise available there could be concentrated in the central nucleus where sires could be bred from screened females to give back to contributors.

ARTIFICIAL INSEMINATION (AI) AND EMBRYO TRANSFER (ET)

AI has been used long enough in farm animals now to be accepted as a very powerful tool for spreading genetic merit in a population. It has been most clearly demonstrated in dairy cattle. Its impact is simply to broaden the base of the hierarchy and flatten the base of the triangle shown in fig. 32 so that fewer sires are spread over a wider base. It is also realised that AI has an enormous power for good or evil in a population so everyone is concerned that the best males only are used. This means that the most efficient methods of identifying them are found so that breeders' commercial needs and profitability are given top priority.

Concern arises periodically about the power that large AI organisations have over decisions on sires and whether they are so record-conscious (or biased) that they neglect aspects of conformation and type. These arguments will continue as long as AI organisations have to be profit-motivated and competition exists between them. The international demand for semen from top sires of all farm animals will grow rapidly – hence the responsibility on breeders to improve them will also increase. AI has allowed very large selection pressures to be used in the drive for maximum genetic improvement and a typical example would be in dairy sires where the total genetic gain was obtained from three different sources as follows:

Sources of total genetic gain (New Zealand data):

Selection among bull mothers	25%
Selection among young bulls	70%
Selection among cows to breed heifer replacements	5%
Total genetic gain	100%

Most gain (70%) comes from selection among the team of young bulls which are bred from about 2–5% of the best dams in the population followed by selection among bull mothers.

Embryo transfer (ET) is where embryos from a particular mating are increased by removing them from their dam soon after conception and placing them in other recipient dams. If super-ovulation is used, the offspring produced by ET are greatly increased.

ET is used to produce offspring from older dams and to produce more top females to become mothers of sires for future progeny testing. Surplus embryos can be frozen for future use or sale. Using older dams will increase generation interval and care is needed to control inbreeding when top related animals may be subsequently mated. The sexing of embryos will add greatly to the value of the technique.

TESTS AND TRIALS

The breeder or commercial purchaser of stock is concerned all the time with comparisons among animals. The whole stock-selling business is based on this concept because as soon as words like superior, top quality, good or bad, etc. are used the reply should be: 'compared to what?'. These salutations of merit are not always based on valid comparisons, so in looking at comparative trials, it is most important to study the details of the trial as these are usually vital in evaluating the results obtained. The farmer has special problems to consider and these are usually concerned with whether the results would apply on his farm under his system of management. Some typical questions are these:

* Were the animals used typical of the breed or the class of stock he was running?
* Was the feeding and management system used typical of the commercial challenge the stock would get on his farm?
* Could he get access to the same type of sires as used in the trial?
* Were the trials run over a long enough period to cover both good and bad seasons?
* Were all the animals in the trial bred in the trial environment, or were they bought in?

The point to stress is that breeders should be aware of these aspects and should seek information from the appropriate advisory authorities before making decisions based on trials.

RANDOMISATION

The aim in randomisation is simply to remove bias in groups of animals or make them as equal as possible. So in progeny testing, for example, the original females should be divided at random for mating to each sire to be tested. Also, if all the progeny of a sire cannot be tested, then again there is a need for the random sampling of those available.

There are a number of ways to do this:

* If the animals are individually identified (e.g. by tags) then allocate them to groups in the office using a table of random numbers. These can be found in books of statistical and mathematical tables (see Fisher and Yates[28]) or Appendix II. The last digits in a telephone directory can also be used. Each animal is allocated to the group in the order of the random numbers.

* The individual numbers of all the animals can be written on small tickets and then they are put into a hat or a box. After shaking, the tickets are drawn out and allocated to each sire group.

* In the stock yards, if the stock all come in as one age group, then they can be drafted-off depending on the facilities. If three groups are needed and there is a three-way drafting system, then the animals are taken off in the order of 1, 2, 3, and 1, 2, 3, etc.

    If seven groups are needed for example, the method is to draft three ways as before but the first draft will take off groups 1 and 2, and 3, 4, 5, 6, 7 will go into another group. The mob is then run through again to split the mixed group into 3 and 4, and (5, 6, 7). One more draft will then split 5, 6, and 7 into three separate groups.

An important practical point is to make sure that each group is drawn from the whole mob, so groups 1 and 2, for example, are drawn proportionately from the whole mob as it goes through the draft. The greatest risk is that some animals that come into the yards last and stay at the back of the mob would not be truly sampled, and they would all end up in the final group. The chances are very high that they would be an atypical group, e.g. a lower social order, or older and perhaps with some sick animals among them. Mixing up the mob periodically is good practice by walking among them before starting to draft.

Randomisation into groups is best done within age if possible, and if there is a wide difference in size or weight among the animals it could be done within these groups also. Where breeding indexes are known, randomisation should be done so that each sire to be tested ends up with dams of similar index in each group.

PRE-TEST ENVIRONMENTAL PROBLEMS
A major problem when comparing animals is to find out what happened to the animals prior to the test. This is referred to as the 'pre-test environment' and deals with the problem of confusing genetic assessments during the test with environmental influences that happened before the test. This is seen particularly in performance tests

of males where the test-period starts after weaning, and such environmental variables as age of dam, age of animal itself, litter size in which born, milk yield of dam, etc. are all confusing the true genetic assessment for growth. Many breeders want the maternal traits included in the animal on test, so they feel a high weaning weight is important to show that the animal had a good dam.

There seems little chance of finding solutions for all these points so that all are satisifed. It seems that the only way would be to start comparisons at birth so that the animals on test were artificially reared. If natural rearing were required, the dams of the animals for testing would have to be run together – probably from early pregnancy – so a simple performance test would end up as an enormous operation in terms of costs and facilities.

Compensatory growth also has to be considered because what happens in the test can be greatly influenced by the pre-test environment. For example, a bull that had a poor dam and had run on hard country would probably respond better to the good conditions in a central performance test than a bull from a very good farm that had been super-fed before the test. Often a 'settling-in' period at the start of the test has been tried to allow for these compensations to sort themselves out. This can rarely be achieved and many view the whole test period as a 'settling-in' period. Even this is not adequate as some animals will never get over the effects of their early pre-test environment.

Table 15 **Some examples of the causes of variation in farm animal traits expressed as the percentage of variation controlled by each cause**

Cause of variation	Calf weaning wt (Angus & Herford)	Calf yearling wt (Angus & Hereford)	Lamb weaning wt (Romney)
Years (seasons)	5	9	5
Age of dam	10	3	2
Sex	7	11	5
Birthday	23	14	15
Type of birth and rearing	—	—	15
	45	37	42
Other causes (?)	55	63	58
	100	100	100

CORRECTION FACTORS

To make valid genetic comparisons between animals, it is necessary to try and remove the bias caused by the main environmental factors present. This is done by using correction factors. Breeders often find the explanation of correction factors difficult although they recognise the need for them. The importance of different environmental factors can be seen from some approximate causes of variation in the weaning and yearling weight of beef cattle farmed under pastoral conditions, and the weaning weight of sheep (table 15).

The variation remaining after these sources of variation have been removed is that due to genetic differences (i.e. the animal's Breeding Value) plus a complex of unexplained environmental differences. The aim is to balance up the animals before comparison so that they are compared on the basis of all being born in the one year, from the same age of dam (a mature dam), of the same sex (a male) and all born on the same day. In sheep it is necessary to add in corrections for birth and rearing rank, i.e. to have been born and reared as a single. An example of a calculation for sheep is shown in table 16.

Table 16   **Example of a correction factor calculation in sheep**

Lamb no.	Birth/rearing rank	Age of dam	Sex	Age of lamb (days)	Actual WWT	Adj. WWT
250	2/2	2	R	+1	20	25.3
	(+4.2)	(+1.3)	(–)	(–0.17)	(+5.33)	
251	2/2	2	R	+1	17	22.3
	(+4.2)	(+1.3)	(–)	(–0.17)	(+5.33)	
120	2/1	3	E	–2	25	29.5
	(+2.0)	(+0.2)	(+2.0)	(+0.34)	(+2.54)	
121	dead	—	—	—	—	—
86	1/1	3	E	–2	22	24.5
	(–)	(+0.2)	(+2.0)	(+0.34)	(+2.54)	

Here lambs 250 and 251 are twins reared as twins so each one receives 4.2 kg to make each of them equal to a single reared as a single. Their dam is a 2-year-old (2-tooth) hence each lamb gets a further bonus of 1.3 kg each. The lambs were one day (+1) older than the average of the flock so it loses 0.17 kg for that. Thus to the actual weight of each lamb is added (4.2+1.3)–0.17=5.33 kg. Note that nothing is added for sex as these are ram lambs. The lamb 120 was born a twin but its co-twin 121 died so 120 was reared as a single. It receives +2.0 kg for being born a twin, +0.2 kg because it is out of a 3-year-old ewe, +2.0 kg because it is a ewe lamb, +0.34 kg because it is two days (—2) younger than average. All this adds up to (2.0+0.2+2.0+0.34)=4.54 kg which when added to the actual weight of 25 kg gives an adjusted weight of 29.5 kg. Finally, lamb 86 is a single ewe lamb which receives +0.2 kg for its age-of-dam correction, +2.0 kg for its sex and +0.34 kg for its age correction. This adds up to (0.2+2.0+0.34)=2.54 which when added to the weaning weight of 22 kg gives a corrected weight of 24.5 kg.

These correction factors are computed from various research trials and large amounts of data that have accumulated in large breeding programmes.

Discussion and argument among scientists and breeders usually centres around how these correction factors should be calculated and how applicable they are to specific flocks and herds. This is especially the case in national recording schemes. Correction factors can either be *additive* where a definite amount of weight, for example, is added to the animal's weight, or they can be *multiplicative* where a proportion of the animal's weight is added on. A lot of discussion usually occurs over which of these two is most appropriate. The difference between the two is described in figs 35 and 36.

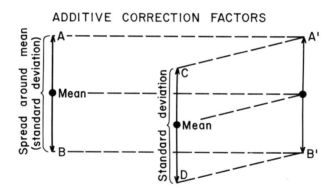

ADDITIVE CORRECTION FACTORS

Fig. 35

In fig. 35 the line AB represents the performance of the standard animal to which the others have to be corrected. The mean is shown as a dot and the spread around it (i.e. the standard deviation) is the line AB. Another animal CD has a lower mean performance for some environmental reason that has to be corrected for. Note that the spread CD is the same as in AB. The task here is to correct CD up to A'B'. This is done by an additive correction factor that adds on the difference between the two means.

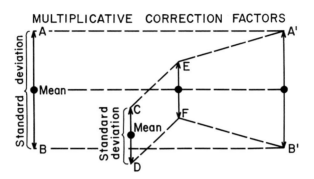

Fig. 36

In fig. 36 things are different. Here the performance of CD has both a lower mean and less spread (low standard deviation) so moving it upwards to EF is not sufficient. It needs a multiplicative correction to widen out EF to A'B' which is then equal to AB. This multiplicative method thus increases the variation and the mean.

COSTS AND BENEFITS OF GENETIC IMPROVEMENT

Breeding improvements are noted for being long-term and hence generally slow to yield a financial return. Often there is a large initial expense in, say, buying stock and setting up the programme followed by a long wait before the 'pay-off' starts. Nevertheless, breeders must face the challenge of having to account for their plans and to do this the technique of discounted-cash-flow accounting has been developed.

This can be explained in an example:

Assume that the money invested in a programme is going to yield 10% return per year, then this becomes a simple calculation of compound interest. For example, for 100 units of currency (pounds, dollars, etc.):

Present value   = 100 units
Value one year ahead   = 100 + (10% of 100) = 100 + 10 = 110
Value two years ahead = 110 + (10% of 110) = 110 + 11 = 121
Value three years ahead = 121 + (10% of 121) = 121 + 12 = 132
and so on.

To calculate the present value of money earned from the programme in future sale of stock etc., a reverse calculation of compound interest is used. Thus 100 units of currency earned in the future is now worth:

Money earned 1 year ahead = 100 – (10% of 100) = 100 – 10 = 90 now
Money earned 2 years ahead = 90 – (10% of 90) = 90 – 9 = 81 now
Money earned 3 years ahead = 81 – (10% of 81) = 81 – 8 = 73 now

Hence by this procedure, all returns and expenses made in different years can be reflected back to the base year and by adding them up an aggregate profit can be calculated in any one year – at current values. For further reading see Bowman.[15]

V Further Practical Advice

Weighing animals

In animal breeding programmes, more time is spent weighing animals and recording live-weight data than in any other practical task. There are still far too many situations where these operations are tiring, frustrating and at times even dangerous.

The following sections are presented for consideration with the aim of increasing the efficiency of the operation.

SCALES

The types and limitations of the scales should be recognised. Scales are generally spring balance (clock face), hydraulic or electronic. Spring balance scales are commonly used for sheep, pigs and calves where a crate to restrain the animal is suspended on the balance. They suffer from the disadvantage that any movement of the animal vibrates the crate and the scale pointer, so patience is needed until the animal remains still or skill is needed to take a visual average reading as the pointer vibrates. This latter situation greatly limits accuracy if this is demanded.

Hydraulic scales are where the weight of the animal forces a fluid up a graduated tube and the level of the fluid indicates the weight.

Hydraulic and spring balance scales are generally similar in accuracy and speed of operation (about 300–500 sheep/hour, 150–175 quiet cattle/hour, or 60 pigs/hour).

Electronic scales are increasingly found in use with data loggers used to record data directly from the scales. Electronic identification is the other development needed to completely automate the system. Electronic weighing units can be used for a wide range of weights such as fleeces, sheep, cattle, and even larger items such as wool bales or groups of animals.

CARE OF SCALES

Scales need little maintenance if kept clean dry and lubricated. However, they are often neglected on farms, mainly through being left unprotected from the weather when not in use. Special care is needed during transport, especially to avoid vibration of knife edges.

TESTING ACCURACY

It is important to know how accurate the scales are and this is best done by using test weights or blocks of concrete or steel of known weights. The accuracy of the scales should be tested over the whole range of weights to be recorded as the error may vary. The recorded weight may also vary depending on where the weight is placed in the crate. This should be tested, as agitated animals rarely stand evenly or still on a scale. Knowing your own weight accurately in working clothes is useful as a test weight. It pays to jump around on the scales to make sure the scale settles at the correct weight each time. The same person should read the scales throughout the operation if possible.

LOCATION AND SETTING UP OF SCALES

This is usually the key to an efficient operation. No amount of extra labour will compensate for a good handling set-up.

A good lead-up to the weighing facility is vital. This should be a race or passage wide enough for animals to move along freely but not be able to turn around and block the flow.

The sides can be open rails or close-boarded depending on the species. Bends in races encourage the flow of stock and the scale may even be put at an angle to the lead-in race to encourage this flow. Decoy animals, especially with sheep are useful and a race which leads stock up hill to the scales may help, especially for sheep, goats and young calves.

For small farm animals (sheep, goats, pigs) hock bars approximately 20 cm above the ground and 1 m apart can be used to stop animals backing along the race when they see people at the scale.

The scales should be set up on a level surface in a position where there is plenty of light, shelter and shade depending on the climate. Check at regular intervals to see that mud, dung or stones have not built up under the scales.

OPERATION

Weighing is usually a noisy operation and this is the first thing to try and reduce for the sake of the people, the animals and the accuracy of the data collected. Noise leads to misread tags, misheard numbers, frustration, exhaustion and bad tempers! Here are some suggestions:

* Make sure the recording person is comfortable and has a modern chair to sit on rather than the oldest chair on the farm. Excess heat, dust, cold and draughts can greatly reduce concentration and hence the accuracy of data. The recording person can even be some distance from the weighing if radio microphones are used.

* Tie up all dogs that are not needed, especially when they are barking, and only use dogs that are under complete command.

* Avoid the use of rattles or other noise-making devices. Tell stockmen to keep the noise to a minimum.

* Do not allow stockmen to poke animals with sticks or other goads as the stock walk up to the scale.

* Oil all the doors and gates around the scale and approach races to eliminate squeaks and bangs.

* Put rubber (e.g. bicycle-tyre inner tubing) on any metal-to-metal surfaces to reduce banging.

* The position of the person reading tags and the person writing down weights is very important. They should be close together and directly opposite each other so that information is heard clearly.

* The person reading tags and weights should have a clear voice with good diction. Avoid people who permanently work with a pipe or cigarette in their mouths.

* Voice pitch is important. Choose someone with a pitch lower or higher than the common noises (bangs and thumps) around the scales. High pitched voices are best.

* Develop a routine where you close gates and doors quietly and avoid bangs which coincide with, and hence drown out, part of a spoken number or weight.

* The tag reader should say each digit for maximum clarity, e.g. for number 167, say one-six-seven, and not one hundred and sixty seven.

* Make the recording person, or anyone who receives the information, repeat it exactly as heard, back again to the reader, e.g. one-six-seven.

* Choose a person with good eyesight and if they wear spectacles, make sure they bring them along.

* Avoid at all costs people who have a known habit of transposing numbers or parts of numbers. They see the written number one-six-seven, but record it as one-seven-six.

* General conversation should be avoided and people who must converse should move away from where the recording is taking place as this is a great distraction.

* Systems which cause stress and frighten animals should be completely re-designed. This is probably most necessary with pigs which make deafening squeals when frightened. Animals are especially frightened of noisy weighing crates that vibrate excessively, and where their ears have to be held tightly or twisted to read tags. Tags should be visible without excessive manhandling of the animal.

Weighing wool

Recording greasy fleece weights in sheep selection programmes has not yet been as fully automated as, say, milk yield in dairy cattle. Research continues into automated shearing where lasers and electronics may hold the keys to this.

At present, problems usually arise because the recording systems cannot cope with a team of good shearers with an output of, say, 250–350 mature sheep in a 9-hour day. Usually the fleeces cannot be cleared away from each shearer, be weighed, sorted and then stacked or pressed quickly enough by the staff. As a result, heaps of unweighed wool with unknown animal identities become stacked up around the operation leading to great stress and panic among the staff. The shearers often see this as a big joke and shear even faster! It is important to let the shearers see that you are running an efficient operation that will not hold them up and hence reduce their earnings. Here are some points to consider:

* Have plenty of scales available so that the people picking up fleeces don't have to wait to have them weighed: one set of scales for two or three shearers is adequate in a good system, but two sets of scales will be needed for four shearers.

* The scales usually have a tray on them where the fleece is placed along with the belly and pieces. Have a number of these trays available, one for each shearer and make sure they are all tared to the same weight.

* Place the scales so that they do not restrict the movement of the people around the operation and yet the recorders do not have to walk too far to weigh their fleeces.

* Don't let the shearers read the sheep's tag numbers; this is a

specialist job and should be given to a reliable person with good eye-sight and who can write clearly.

* A system of brightly coloured labels or tickets is the most simple to start with, because the ticket stays with the fleece. The sheep's tag number is read and then written on the ticket. This is put near the shearer or on the belly wool if it comes off first and separately, as in the Bowen method.

When the fleece is picked up, the ticket is put on top of the wool when it is weighed. The weight is written on the ticket, the fleece goes for sorting and pressing, and the ticket goes to the recording person or into a box to be written up later.

Many modifications to this sytem are in operation. For sheep without permanent tags, a temporary tag system can be used or the animal held until the fleece is weighed and the 'keep or cull' decision made. This latter method will slow up the work rate but may be suitable in small operations.

All the above remarks on weighing wool would also apply to systems of shearing goats for mohair or cashmere fibre.

Mating records

In all breeding programmes it is vital that the parentage of animals is known accurately. In practice, parentage sometimes has to be assumed as actual mating was not seen or was recorded wrongly. Males may be joined with females (that is put in the same pen or field) but the females can sometimes be mated by other males. Typical situations are where stray rams run loose on open grazings or feral bucks or bulls come in from the adjacent bush or desert areas. Broken fences, broken gates, defective catches, doors left open and other accidents or errors can ruin the best of breeding plans and set them back for years. Mating time is a time for constant vigilance by stockmen.

The main area of error on most farms however is where sires are changed during the joining period. This may be because of injury or it may be to reduce the risk of having an infertile sire.

Errors in sire identity occur mainly because people do not appreciate the variation possible in gestation length. In sheep for example with average gestation of 147 days, it is wrong to assume that you can count 147 days back from the birth day of the lamb, see which ram was with the ewe, and then call that ram the sire of the lamb! Gestation length may average 147 days but a common range is from 140–154 days. Thus

an accurate pedigree cannot be guaranteed within 3–4 days either side
of a ram-change day. This same principle applies to all farm animals so
the range in gestation length is as important to know as the mean.

Here are some practical suggestions:

* After the females are chosen for a mating group, either by selection
  or randomisation (see page 135), it is good practice to give each
  group a distinctive temporary mark (not on the rump). This is to
  allow rapid sorting if groups become accidentally mixed. It is also
  useful to give the male the same group mark (and much more
  distinctive) so that he can be recognised quickly from a distance,
  and if in the wrong group he can be removed quickly.
* Mating groups should be checked regularly, especially during the
  early part of mating, and at the peak oestrus time. Also, as mating
  progresses, males often become bored with their own group and
  start fighting their neighbouring group's male through the fence,
  being all too anxious to get through and help him out.
* Use specially-prepared recording sheets to suit your needs. One
  suggestion is as follows:

Female no.	Male no.		Cycle				Notes
		1	2	3	4	5	
123/74	205/74	√	√	√	√	√	Barren
221/74	307/74	√					
225/74	~~525/74~~	√	√	-----			Sir change 20/10
	500/74			√			

Individual females' numbers are written down in numerical order
within age groups prior to mating so that they can be quickly
located in the field. Recording the mating by each sire in each cycle
is only possible if animals are hand mated, artificially inseminated
or the male has a harness and marks the female with a different
colour at mating. Care is needed to make sure these marking
devices work and in interpreting the marks made. Some may not
be proper mating marks but happen during false mounts or in the
yards when animals are crowded together.

* Another method is to have a separate record sheet for each male
  with his mates listed in numerical order. With many males in use
  this results in a lot of paper.

* A method to reduce the quantity of paper is to have all the males and their mating groups on one sheet as follows:

Male no. 25/79	Male no. 63/79	Male no. 75/79	Male no. 35/75·
2/74	5/74	3/74	8/74
15/74	25/74	30/74	20/74
3/75	2/75	60/75	56/75

Again, this should be made up in the office before going to the field.

In the field a suggested practice is to use a:

* Tick to show that each female was checked as being present at joining when the male went out;
* Pencil line through the female's number when mated;
* Coloured line through the female number when remated;
* Circle around the number to show female was not remated.

Coloured pens improve the speed and accuracy of recording. Double spacing is essential to allow room for brief comments about animals and alterations.

Birth records

Recording at birth is the other critical time to establish pedigrees. Here the birth process is often observed by stockmen so identification of offspring to their dam should be highly accurate. Accuracy is certainly high where females give birth indoors or in individual pens, but with species that have litters and give birth outdoors, mis-mothering may be common and errors can occur. An example is in sheep where 17% errors in correct dam/lamb identity have been observed. The most likely causes of error at birth are wrong dam/offspring identity, incorrect tagging and then incorrect entry in the record book.

Some suggestions to reduce these problems where animals are not supervised and give birth outside are these:

* Avoid overcrowding caused by high stocking rates.
* Recognise any areas which are preferred by the animals to give birth such as sheltered areas or spots away from disturbance.

Management may have to be altered to avoid over-use of these areas with the resultant mis-mothering.

* Consider drafting animals into sire groups before birth. This will help to record the correct sire identity but will not guarantee correct dam/offspring recording. This practice can also lead to environmental bias when sire groups are subsequently ranked. If it is used, each sire's group should be drafted out only a few days before birth, and mixed up again 3–4 days after each female gives birth.

* Avoid unnecessary disturbance during the birth period and the recording technique should cause as little disturbance as possible. Animals should be accustomed to the basic disturbances, such as dogs, people, and vehicles, before the birth season begins. Stock which have been on rotational grazing systems should be set stocked at least a week before birth, to break them from the habit of seeing people or hearing a vehicle and assuming it is time to move.

* Recognise the importance of the animal's birth site in the correct bonding of offspring to dam. Dams and offspring need time to allow for good bonding before they are moved.

* Develop a system of temporary marks until bonding is complete. Often daybreak is the best time to identify offspring with their dams before the animals move and become temporarily mixed. Small coloured spot marks on various places on the body can be made into a code which will last until permanent tagging is done later. Remember that even animals a few days old can move quickly and catching them can cause even greater disturbance and mis-mothering, so the timing of permanent tagging must be given special consideration.

* In some species, like cattle, the offspring do not always follow their dam as do sheep and goats but often lie in groups of their own. Sorting out dam/offspring pairs needs special care.

* If the recording person is not sure about a dam/offspring bond, then it pays to be honest and record this. Use a simple code such as 1 = observed birth and certain; 2 = birth not seen but no reason to doubt; 3 = doubtful. The value of this is during final selection of sires perhaps years later when a 'doubtful' observation could lead to the animal not being used as a top stud sire or for artificial insemination. It is at this stage where the honesty of field recorders will pay greatest dividends.

Recording information – general points

The prime aim is to collect information as quickly as possible and with a high degree of accuracy. This will result in the lowest cost of operation. Errors take time to sort out and inevitably mean loss of information and increased processing costs. Here are some suggestions:

* Always have the animal numbers pre-listed. This reduces the chance of numbers being copied down wrongly. Printed check lists can be used – these are lists of numbers 1 to 1000 with spaces beside each to write information in.

* Place the recording sheets on a firm base for writing and use a system of clipping or tying them down to prevent them flapping in the wind.

* If possible, avoid turning pages in the recording sheets. This can be done by laying them out across a large board.

* If many pages are necessary, fix numbered tabs which protrude from the side of the pages to allow rapid turning and finding the tag numbers.

* Use good quality paper which is lightweight but strong. If possible use paper which is waterproof. A pencil with a medium-soft lead is better than ink or ballpoint. A pencil will nearly always write on damp and greasy paper.

* Develop a technique to protect the records from the weather such as keeping them inside a clear plastic bag, a small towel to dry the hands before writing or a piece of card to rest a wet hand on when writing.

* Carry spare pencils, pencil sharpeners and pens, and if spectacles are needed a system of keeping them readily available and safe in the field or yards is important.

* An easy to see tally method is useful. Rather than writing tallies as ||| ||||| ||| ||| and then having to count all the strokes at the end (when miscounts are highly likely), use a system of marking fives ‖‖ ‖‖ ||| to make final counting quicker and less prone to error.

* Unnecessary copying of data from one sheet to another should always be avoided. An example is where some people have a field list which they complete at weighing, then they copy this into a 'good' book for the office and then this is copied again on to permanent record cards or a computer input! The system should be organised so that the computer input forms can be handled in

the field and the information be recorded on them directly. Every time you transcribe a list of figures, you will make errors so it should be avoided. Smudged but accurate computer inputs are preferred by punch operators to clean ones full of errors.

* Good clear figures and writing are essential and people who cannot meet these needs should not be asked to record. Data punch operators do not have time to decipher writing – this will result in errors and increased charges.
* If data has to be transcribed, this should be done immediately after it is collected in case the original gets lost.
* Someone should be responsible for the safety of the records in the field so that the information is not lost, walked on by animals, covered in mud or even eaten. Pigs are notorious for eating paper records!
* Advances in electronics where data go directly from scale into a processing unit are proceeding and hopefully will avoid many of the problems mentioned above. However, this requires more capital investment and may take some time to reach all situations.

It has been noted that with the automated systems developed so far, if people are involved, they need something physical to do, such as opening a gate or pressing a button, to help them maintain concentration. Otherwise they day-dream and make errors.

COMMON CAUSES OF ERROR: A CHECK LIST

* Misread tags – the result of insufficient time and care when reading, poor eyesight without the use of glasses or poor tags where numbers are not clear or tags are badly positioned on the animal.
* Numbers wrongly written down – transcribed numbers. Read them out aloud when you read them, and read them aloud again when you write them down. Be vigilant if you have this problem.
* Numbers misheard by people when others do not speak clearly. Always repeat a number out aloud back to the person who spoke it to you.
* Failure to have a set recording routine and failing to stick to it when there is one. Unless you do, things will be missed out.
* A system where there is no opportunity for cross checking to spot errors before the data is processed.
* Using old, worn, dirty and broken tags when trying to save money. This is false economy and frequently results in duplicated numbers and utter confusion.

* Writing which is not clear and done with a blunt pencil or blotchy ballpoint.
* Assigning offspring to dams without checking that the dam has actually given birth or finished giving birth – she may have stolen the offspring from another dam.

Publicity and promotion

Breeders are sometimes disappointed to find that after years of carrying out a breeding plan to increase genetic gains, this has not resulted in large extra profits through the sale of their stock – and they voice their feelings strongly to geneticists and advisers!

The breeders point out that their programme has meant extra costs such as paying recording scheme fees, employing extra labour, making extra improvements around the farm and so on. They add that buyers have not automatically rushed to buy stock, those that came were not interested in the paper records and the biggest problem of all, none of them wanted to pay the higher prices necessary to cover the extra costs!

The first approach to help breeders is to ask them to consider what they have done to help themselves! Have they told anyone outside what they have been doing in their programme over the years, how good it is and how buying from their flock or herd will benefit the client? Many breeders reply to this question by saying that promotion and publicity is not their job as they have no expertise in the area – they leave it to their breed association, auctioneer, stock and station company, bank manager and the like. Some even say that this should not be necessary as the stock should sell themselves if they are any good – satisfied customers will always come back!

This approach is dangerous and wrong. The main point to appreciate is that the most informed and committed promoter of a flock or herd has got to be the owner, regardless of whoever else he may use to help him. Publicity and promotion must start with the breeder and there is a wide range of the things which can be done. Genetic gain is a very marketable factor once it is identified and packaged correctly for the buyer – there are plenty of commercial farmers keen to obtain value for money spent on genetic improvement for their flocks and herds.

PROMOTION: FIRST CONSIDERATIONS

There are a number of basic questions to ask before any promotional plans are drawn up. These are:

* Who is your promotional message for? Is it for other pedigree stud breeders, for commercial farmers, for overseas buyers, for servicing people who lend money and so on? You must clearly identify your target otherwise money will be wasted and credibility may be lost.
* What is your message? Tell someone verbally what you are offering and why it is so good, then write those spoken words down on paper. Keep it brief and underline the key words. If your message is brief, then you will remember it, and then you and your staff will be able to quote it in conversation or writing whenever they need to. It becomes like a school motto they all believe in.
* What result do you want from your message? Do you want to sell more sires, surplus females, semen or frozen embryos or would you be satisfied with people knowing you were a 'switched-on' breeder or a nice person? You have to be very clear in your objectives.
* Is the method you are using for promotion the most appropriate for the message? Perhaps you have planned a field day and a brochure would have done. You should question the plan you have formulated – get other people's honest views on it too.

All of these questions are aimed at making sure you get a return for the money you are investing in promotion. Your livestock will eventually have to pay the bill.

PROMOTION METHODS

ADVERTISEMENTS
These are usually done for you by agents or the advertising departments of magazines and newspapers whose expertise can be used in layout. You should provide them with a good clear message or slogan and a selection of top quality black and white photographs. It is with photographs where problems regularly occur – usually the only photos available are in colour and they are taken with snapshot cameras and have poor definition. It pays to consult a professional photographer to discuss your needs.

BROCHURES
These are always useful as they can be given away by hand, mailed to clients or prospective clients, or left at various places for people to collect on a self-help basis.

Any printer will help you to design one but it is best to make a mock-up of the size and shape you want and sketch out on it the message you want, its position in the brochure and the photos to be used.

If it is a folding brochure, the way it opens and is read can be very important. Your message should be tailored to fit this. Remember that people often read things the opposite way to the conventional one.

CALENDARS

These can be produced by a commercial printer who will provide the photograph or pictures together with your message. The pictures may not be very relevant to your flock or herd so the answer is to provide your own. Again, high quality is essential.

Most of the cost is in making the plate for the picture so you need to have a fairly large circulation and a mailing list. Cost of distribution must also be considered.

NEWSLETTERS

These are fairly cheap to produce and can be done as personal letters. It is important however that the paper is good quality, the printing is clear, the layout is good and the photographs, if included, are high quality. A circulation list is needed and methods of distribution as well as frequency of publication should be decided.

FIELD DAYS AND FARM WALKS

These are always successful and rewarding, as you can meet your clients and their friends on a face-to-face basis. Here are some keys to success.

* Advertise it well. Send out personal invitations to your clients and ask them to bring along anyone interested. Send an invitation to the local press, radio and television outlets, bank managers, servicing agencies and so on. Spend some money on a good-quality invitation with a means of obtaining a reply from them to calculate the catering needs.
* Include a good clear map on the invitation showing the way to get to your farm.
* Choose a time of year which suits the majority of your clients.
* Have the farm well signposted at critical road junctions and long before your gateway. Mark the entrance clearly and then have notices for the main facilities such as parking, refreshments, toilets, demonstrations, information, telephone and so on.

* Make any helpers wear distinctive dress such as white coats. This helps anyone with a query to find someone who will help them. All these people should wear a name tag.

* Invite visitors to wear a name tag. This usually embarrasses them but stress that it is for your benefit to get to know them.

* Catering: Whatever meals are needed make sure there is plenty of food, not only for the main lunch but also for those people who may come very early or leave late. Special attention is needed for those who have come a long way. People form important impressions from your catering.

* Toilet and washing facilities should be clean, fully serviced and clearly signposted. This is a very important part of the 'general impression' people will form about your operation.

* Having the stock in top-class order goes without saying. Any stock which are not looking their best for whatever reason should be put out of sight.

* Have good labels signwritten for the different groups of animals along with any performance information.

* Have plenty of copies available of any performance information you wish people to take away, along with your contact name and addresses.

* A good introduction and welcome at the start of the day sets the tone, and this is the stage at which to give any formal presentation of information that may require charts, slides or videos. Make sure there is plenty of seating available because it would be a mistake to tire the visitors at the beginning of the day by having them stand throughout the presentation. Talks often go on longer than intended, especially if questions are invited. Hay bales are useful as seats.

* If it is difficult for all the people to see the stock all at the same time, then it may be necessary to bring the stock to the people. This will require much more elaborate facilities, almost like an auction ring. Make sure the stock have gone through the facilities, but remember they are likely to be frightened by the crowd and especially by any applause during votes of thanks. Make the facilities strong enough to avoid accidents.

* Have someone trained in first aid if the crowds are large. Also have fire-fighting facilities checked and available.

* Have a general tidy up around the farm and make any repairs necessary to the road, entrance, gates, steps, fences, gate catches, pens etc. Apart from safety, these things are vital in giving good general impressions.

DISPLAYS

These are always useful as you can use them in your talk if you have a group of visitors or clients, or they can be used as a general backdrop in stud sheds, offices, foyers, shop windows, banks, sale yards and so on.

Before you plan a display again decide the objective. Here are some objectives which you can put in your order of priority before planning. Aims of displays are to 'teach, motivate, promote, offer a service, show new developments, create awareness, improve your image, gain recognition'.

Displays that have good effective messages usually have the following things. They are simple, have a purpose, arouse interest, are topical, stimulate thought and finally, bring about some action. It is useful to look at displays at shows and exhibitions to check these points – especially at those with large crowds and also very small crowds around them.

Remember that where people can walk past displays, you have about 90 seconds to attract their attention and give them your message before they move on. If your display does not stop people then there is little chance that they will read or absorb your message.

Thus the main aim in a display is to gain attention and this can best be done by different forms of 'action'. Examples are movement, heated discussion, noise, flashing lights, smells, colour, humour, give-away items. This does not mean that displays need to be sophisticated; again the top priority is simplicity, and the worst fault and the most common one, is where too much information is provided. This is an easy trap to fall into especially when there are large amounts of performance information available on animals. The solution is to keep the display simple and give any necessary statistics in an attractive handout.

Large, good-quality photographs are an important part of good displays. As photographs are usually expensive they need careful mounting and protection during transit. Indeed the display should be made with transport as a major priority as it may have to be carried folded in a car boot and handled and assembled quickly by one person. Assembly instruction should be clearly attached to the display.

Films and videotapes are regularly used in displays and these are very effective in attracting attention. They should be professionally made and good facilities such as plenty of room and seating may be needed for their viewing. Back-up equipment or maintenance people should be available in case of breakdown.

Probably the most important area around a display is where people can sit down, rest, leave their bags in safety, have some refreshments and talk. It is here where the breeder can use face-to-face communication to get the message over.

GIVE-AWAYS

These are used by some breeders and include such items as pens, pencils, key rings, notebooks and such like which have the breeder's name, address and telephone number on them. The breeder's objective or slogan may also be included. These items are always acceptable to clients but are best used as part of some other positive way of getting a message across.

VISITS TO CLIENTS

Face-to-face communication is without doubt the most efficient way for a breeder to get a message across to clients. The fact that the breeder visits the client gives a clear impression that the breeder is interested and cares about the client's problems. It is extremely good public relations but can be very time consuming. It is generally done at quiet times of the year, out of season and is a means of the breeder making sure that next year's orders are known in time to make any alterations to planning.

If clients have any complaints, then at least they can tell the breeder directly. Breeders who visit their clients soon realise also that they have an important role in education. It is very good practice, indeed essential, to make sure that the client's partner (wife or husband) and any interested members of the family are there when technical points are discussed. They are often able to grasp new concepts and carry on the discussion and questioning long after you have gone home. It is here where give-away items may be most successful.

References

1. Lerner, I. M. and Donald, H. P. (1966) *Modern developments in animal breeding.* Academic Press.

2. Hammond, J. ed (1955) *Progress in the physiology of farm animals.* Volumes 1, 2 and 3. Butterworth Scientific Publications.

3. Hammond, J. (1956) *Farm animals. Their breeding, growth and inheritance.* 2nd edn. Edward Arnold.

4. Berg, R. T. and Butterfield, R. M. (1976) *New concepts of cattle growth.* Sydney University Press.

5. Kelly, R. B. (1949) *Sheep dogs.* 3rd edn. Angus & Robertson.

6. Sinnott, E. W., Dunn, L. C. and Dobzhansky, T. (1958) *Principles of genetics.* McGraw-Hill.

7. Strickberger, M. W. (1968) *Genetics.* Macmillan.

8. Winters, L. M. (1948) *Animal breeding.* 4th edn. John Wiley & Sons.

9. Lush, J. L. (1945) *Animal breeding plans.* 3rd edn. Iowa State College Press.

10. Auerbach, Charlotte. (1962) *The science of genetics.* Hutchinson.

11. Carter, C. O. (1962) *Human heredity.* Pelican Books.

12. Hagedoorn, A. L. (1946) *Animal breeding.* 2nd edn. Crosby Lockwood.

13. Wright, S. (1968) *Evolution and genetics of populations. Volume I. Genetic and biometric foundations.* 1st edn. University of Chicago Press.

14. Lerner, I. M. (1968) *Heredity, evolution and society.* W. H. Freeman & Co.

15. Bowman, J. C. (1974) *An introduction to animal breeding.* The Institute of Biology's Studies in Biology No. 46. Edward Arnold.

16. Falconer, D. S. (1960) *Introduction to quantitative genetics.* Oliver & Boyd.

17. Turner, H. N. and Young, S. S. Y. (1969) *Quantitative genetics in sheep breeding.* Macmillan.

18. Kelly, R. B. (1960) *Principles and methods of animal breeding.* Revised edn. 1960. Angus & Robertson.

19. Snecodor, G. W. and Cochran, W. G. (1967) *Statistical methods.* 6th edn. Iowa State University Press.

20. Lerner, I. M. (1958) *The genetic basis of selection.* John Wiley & Sons.

21. Johansson, I. (1961) *Genetic aspects of dairy cattle breeding.* University of Illinois Press.

22. Johansson, I. and Rendel, J. (1968) *Genetics and animal breeding.* W. H. Freeman.

23. Rice, V. A., Andrews, F. N., Warwick, E. J. and Legates, J. E. (1962) *Breeding and improvement of farm animals.* 6th edn. McGraw-Hill.

24. Cundiff, L. V. and Gregory, K. E. (1977) *Beef cattle breeding.* United States Department of Agriculture, Agricultural Research Service AGR 101.

25. Preston, T. R. and Willis, M. B. (1970) *Intensive beef production.* Pergamon Press.

26. Owen, J. B. (1971) *Performance recording in sheep.* Technical Communication No. 20. Commonwealth Bureau of Animal Breeding and Genetics, Edinburgh.

27. Ryder, M. L. and Stephenson, S. K. (1968) *Wool growth.* Academic Press.

28. Fisher, R. A. and Yates, F. (1948) *Statistical tables for biological, agricultural and medical research.* 3rd edn. Oliver & Boyd.

Appendix I

The coefficient of inbreeding

When a population is closed (i.e. no more genetic variation is introduced from outside) and breeding continues at random, then it is inevitable that there is a slow build-up in the level of inbreeding through relatives mating together. The rate at which the resulting heterozygosity is reduced (or conversely the homozygosity increased) is described by Lush's formula[9]:

$$\Delta F = \frac{1}{8M} + \frac{1}{8F}$$

where ΔF = the increase in inbreeding per generation
M = the number of males in the population
F = the number of females in the population.

Thus in a herd of two sires and forty females this means that $(1/16 + 1/320)$ or about 6.6% of the heterozygosity is lost. Generally the males are least in number so the $\frac{1}{8}M$ part of the formula is the most important, and the $\frac{1}{8}F$ part can often be ignored.

The above formula describes the situation in whole populations but when it comes to examination of inbreeding in individual pedigrees, Professor Sewell Wright's formula is generally used.[13] This is as follows:

$$F_X = \Sigma \left[\frac{1}{2}^{ns + nd + 1} (1 + F_A) \right]$$

where
F_X = the coefficient of inbreeding of the animal under study
F_A = the inbreeding level (if inbred) of the common ancestor out of which the line of descent divides
Σ = the sum of
ns = the number of generations from the *sire* to the common ancestor
nd = the number of generations from the *dam* to the common ancestor.

The important points when working out an inbreeding coefficient are these:

(a) Recognise and mark the common ancestors in the pedigree (i.e. the same animal on both the sire and dam's side).
(b) In complex pedigrees draw an arrow diagram to simplify the recognition of the lines of descent from the sire back via the common ancestor to the dam. This is where care is needed to avoid errors.
(c) Remember that although we are concerned with the subject animal of the pedigree, the lines of descent end at the sire and dam. It is because these are related that the subject is inbred. The offspring would not be inbred if the parents were unrelated to each other, even if each parent were itself inbred.

EXAMPLE 1

$$
X \begin{cases} A \begin{bmatrix} \text{Ⓒ} \\ E \end{bmatrix} \\ \\ B \begin{bmatrix} \text{Ⓒ} \\ E \end{bmatrix} \end{cases}
$$

This is a simple pedigree with one common ancestor C. The lines of descent can be drawn as follows:

Thus:

$$ns = A \rightarrow C = 1$$
$$nd = C \rightarrow B = 1$$
$$F_X = [\tfrac{1}{2}^{\,1+1+1} \, (1 + F_A)]$$
$$= [\tfrac{1}{2}^3 \, (1 + 0)]$$
$$= \tfrac{1}{2}^3 \text{ or } 0.125$$
$$= 12.5\%$$

EXAMPLE 2

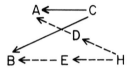

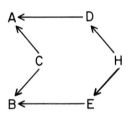

This can be drawn as an arrow diagram as follows:

(arrow diagram: A ← C, B ← E ← H, D)

or like this:

(arrow diagram: A ← D, C, B ← E, H)

* Inbreeding due to C:

$$ns = A \rightarrow C = 1$$
$$nd = C \rightarrow A = 1$$

$$F_X = [\tfrac{1}{2}^{1+1+1} (1 + F_A)]$$
$$= [\tfrac{1}{2}^3 (1 + 0)]$$
$$= \tfrac{1}{2}^3 = 0.125$$

* Inbreeding due to H:

$$ns = A \cdot D \rightarrow H = 2$$
$$nd = H \rightarrow E \rightarrow B = 2$$

$$F_X = [\tfrac{1}{2}^{2+2+1} (1 + F_A)]$$
$$= [\tfrac{1}{2}^5 (1 + 0)]$$
$$= \tfrac{1}{2}^5 = 0.031$$

The total $F_X = 0.125 + 0.031$
$$= 0.156 \text{ (or 15.6\%)}$$

If H had been inbred, say 25%, then the expression $(1 + F_A)$ would have had a value greater than one and the formula would have been:

$$F_X = [\tfrac{1}{2}^5 \times (1 + 0.25)]$$
$$= 0.031 \times 1.25$$
$$= 0.039$$

Note that although F and G appear on both sides of the pedigree, they are ignored as they are the sire and dam of C and appear in the pedigree only via the animal C.

Most texts cover the calculation of the coefficient of inbreeding in detail using many examples. [9, 16, 18]

THE COEFFICIENT OF RELATIONSHIP

This is used to describe how closely related two animals may be and is calculated by another formula. [9, 16, 18] A useful short-cut method to find the relationship between one animal and another is to work out the inbreeding that would result if they were mated together (regardless of their sex) and then double this figure to give the coefficient of relationship.

Appendix II

Random Numbers

```
2387  9003  3951  5695  1284  4761  7118  1196  1741  3791  3405  3132  6682  9493  9864
7359  1250  7036  2916  7562  9299  8910  6713  5173  8617  4222  0244  3045  4923  1740
3200  2876  4591  6695  0574  6376  6141  1322  6031  0193  0469  9160  4010  4583  9834
3100  0586  5833  5593  7181  3078  2430  4682  2133  8413  4987  4854  4680  0229  9913
8661  0584  4736  6834  6189  1441  5331  9766  6544  5230  5949  2036  8082  0606  3178

4954  7400  4954  7117  0237  7064  6177  0984  6178  2096  8250  9478  7569  5211  0513
5885  9373  0797  0721  7942  4146  7649  1836  2440  3209  0461  2579  8590  1568  0050
1725  0128  6589  2347  6147  6176  0417  6131  2618  0851  6151  9509  3355  1625  7762
5157  5315  6433  3768  4298  3244  9535  4655  9440  6473  6120  4048  5861  9464  5673
3633  1840  6108  4079  1830  7645  7773  4482  6216  3263  5098  6899  9692  4889  3542

6574  0410  0160  2679  1092  5442  2134  9247  6234  8205  2031  3945  0590  9626  0714
1976  8918  3074  8923  7351  4249  3023  6261  0927  8754  7477  8929  2211  3346  4145
6206  4905  6043  0356  5689  6199  9079  6739  7273  9824  4758  4739  5890  2846  0069
7289  2986  6212  0041  4736  6766  9046  9858  2510  1166  6701  5710  4045  3597  2550
4411  6853  1427  3536  1752  2445  7212  9468  9614  0003  2685  6069  1826  3801  1485

1771  8774  4379  0649  8263  0764  3463  0838  4789  0123  2636  5327  6190  2069  5763
7737  4709  4312  5161  8522  5531  5418  1520  0200  6081  1915  7577  1337  9076  4156
5535  9287  9313  0926  8240  1669  4225  2803  4097  6608  8171  4265  5743  1713  0488
4550  2475  1818  7282  7269  6411  5460  6982  8760  2424  0120  9632  5201  1804  6200
9937  8467  9686  9763  4066  0743  6217  4017  0190  7878  1483  9544  2788  1500  4177

3293  0687  9872  4024  6307  8863  7500  7949  0209  3984  3498  0002  2441  3713  8691
5030  0240  9748  1479  6095  3383  4622  2443  5652  5660  2252  7708  0464  5117  2232
1049  0281  3037  7882  4752  0255  1674  8044  9224  7117  0739  2631  1349  9098  2631
1919  1603  9597  6428  9601  9909  4608  0614  9578  0245  3889  4567  6506  4011  5885
8793  0099  2130  5397  3912  3167  5911  1011  8387  3421  6660  1260  5016  9015  4034

4555  6161  7844  3157  7850  7732  0591  0476  5068  1927  7525  8369  8273  8708  9045
8507  2588  6043  9980  3962  4872  8887  7635  0067  6136  6916  3587  4026  7721  1012
9231  3270  1216  8498  5081  2571  2131  7121  3811  9900  7319  7788  7057  1032  6147
6050  5962  2913  5418  2334  5471  5763  8046  0931  3381  3939  5473  7728  7135  5541
3844  6694  9373  0776  3926  5111  7224  9064  4785  7563  0355  4898  1204  8669  6770

7898  8002  4418  2747  8079  7993  6863  9542  0849  4531  6955  5826  9971  6233  7887
8640  3204  6906  5719  1116  5982  9532  2422  8333  8828  9002  2680  1928  8532  3600
4431  3453  3070  5239  3168  6490  0275  8443  9984  7503  0263  8086  3372  5454  1599
5868  4764  0158  1225  5558  7840  9394  8126  6974  1561  4765  0758  8717  6979  6306
8214  6959  7775  5844  5149  9173  4558  9107  0453  6119  2915  6585  9670  6580  5202

3137  1170  0345  6099  6352  6074  6142  1898  3657  1924  5625  3556  8178  0103  6107
3490  3349  7010  2045  6123  6271  8981  5274  2183  9820  0957  3988  6747  3508  8914
0483  1041  3095  6028  7633  7984  8941  2598  4127  1487  3123  9720  8455  9499  5291
4991  7841  0264  9480  9545  3517  5701  6513  9644  5639  4982  0201  7013  3858  8432

0485  2823  0417  5787  7517  2305  1324  7674  2205  8107  1273  5851  0509  2403  2318
2185  1493  7751  6292  6392  9115  6708  0980  2518  7780  0180  0041  4175  1633  4543
6680  8525  7608  7492  1072  9013  2222  2655  1528  6854  2126  2954  9422  1264  8398
2711  7863  8449  4594  8509  4112  9334  8538  8800  5761  6052  7879  2268  0647  7176
9378  4934  0532  1703  1961  6017  8518  1789  3513  2545  0222  4235  1426  2712  8698
```

```
0720  6897  7989  3108  6782  2972  3982  1145  9561  6345  0042  5204  4273  2904  7626
2303  9453  7029  7403  7757  0934  4638  5117  2599  4583  9983  6193  4750  7787  6199
9100  3944  9685  9598  7455  0398  0154  7721  0889  7234  7540  1127  4652  6652  5149
8948  7973  9851  6475  8179  1170  0698  9187  6732  1511  3157  7865  0226  7819  2455
5479  2537  3835  0004  3631  0096  0328  2076  1531  9934  5949  2224  3741  1479  6848

7450  6649  2104  5027  8134  3711  6423  5454  1700  9432  9338  1118  0038  2102  2836
1641  0681  5279  6536  8638  1722  8024  1985  5826  0487  3885  0706  5881  5957  8632
5996  0850  4941  6741  8468  0509  1998  6628  5355  9472  3105  4636  3550  3055  3085
7764  6407  1877  6773  5155  4288  5227  3942  8333  8857  3170  7973  0182  1130  7404
8278  2681  3270  0876  6469  3394  4132  6135  6285  0697  8567  2445  7259  7768  0460

1643  2141  5319  9532  2747  8073  3078  1888  5321  1507  0029  4308  1576  6879  3392
2300  7236  9030  6359  1895  1145  9378  4578  5191  3625  5840  1386  9654  3590  6503
1798  1163  4399  7050  5236  0806  0470  1926  7014  5429  0638  9262  8578  1930  0264
9549  6484  5310  2096  8058  0133  0465  5875  1011  8260  7737  4699  5505  4164  2210
2126  3184  9694  6752  8043  7657  7984  8821  8263  5675  4934  1004  2897  2050  9451

5317  8453  8326  2426  1252  8703  5204  4472  7574  9286  8712  2409  7866  1073  0126
4499  9827  7273  9643  4425  8322  9292  3804  4031  2172  1347  6664  4806  4709  3817
4714  8115  8064  5362  5553  4187  1222  3776  0313  9845  1816  5596  9825  4995  1670
4468  4545  7860  6101  8248  8225  8958  6966  4867  5336  3847  9453  6938  2237  4599
2635  5117  2098  0040  3763  0209  3576  4736  6313  4032  3228  6721  1784  9024  1693

3987  5243  6602  3606  9716  3625  4271  1274  6265  4777  1486  2096  7900  9257  4275
9737  2280  0878  8515  9047  0694  6009  2213  4737  7041  7619  6192  3794  5568  6072
3789  1661  7351  4679  9194  2153  5535  8992  4561  1133  9312  9850  5867  3928  6658
0177  7522  6508  5820  5531  5730  0404  5546  8396  0592  1380  4072  6232  6974  1853
6855  3273  4109  7072  3059  6072  3962  5031  0787  2796  8335  0146  1069  6278  4729

0676  0592  0809  0952  9895  3187  2688  8481  1013  0065  4516  4566  5848  8797  3248
3096  7250  1051  1983  4204  5270  9077  5515  2626  7081  0940  9631  4813  0656  4504
4155  4791  2183  0168  9745  8766  7240  2350  8442  8486  5717  9720  7965  3354  0487
4185  9760  1278  9829  8625  0794  8291  3598  3376  9103  6903  3150  1463  3091  3214
5009  2586  4575  2772  8294  5857  5579  5182  5888  1277  8887  7466  9360  9978  2335

6173  7639  3204  6861  8238  9433  0046  8299  0590  9401  3923  2929  8513  8079  7824
6770  2756  5546  8285  8631  5390  8851  8126  6905  4418  3182  7937  0117  7137  6786
6310  1221  2243  9648  8824  5646  0694  5905  5781  3094  5319  0022  8688  3087  9175
6713  5493  4261  2659  4613  4568  7417  9365  3700  7445  2316  0695  6465  0299  8225
9219  2839  4183  8215  0782  8858  4013  7207  5366  9090  6301  3771  6698  2805  5640

5670  0607  3896  9938  9158  2140  4552  4405  2028  1414  2683  4282  0408  8250  3083
6332  9306  3431  2363  9067  6913  1010  7613  1740  2640  8973  8832  2073  8925  8944
5093  2746  6772  4528  4471  6628  5054  0419  7709  0764  3556  8257  5644  9107  9772
1242  0378  4004  9454  7491  0200  7430  0907  2145  8256  4973  3037  8232  4974  3567
7148  6277  4231  7976  2488  2657  3282  1032  5544  6714  6425  6614  3096  6613  2427
```

Reprinted with kind permission of K. J. D. Quinn 1974: Eton Press Ltd; Box 8203, Christchurch, New Zealand.

Glossary

Abortion: expulsion or removal of the foetus from the uterus.

Ad libitum: unrestricted.

Additive: combined effect.

Albinism: complete absence of pigmentation.

Allele: any one of the alternative forms of a gene occupying the same locus on a chromosome.

Amniotic fluid: the fluid around the foetus.

Antibody: defensive substance produced in the animal as a response from invasion by an antigen. Antibodies confer immunity against subsequent re-infection by the same antigen.

Artificial insemination (AI): the technique of collecting the male sperm and inserting it via a pipette into the female reproductive tract.

Artificial selection: selection by man. Opposite to natural selection caused by nature.

Assortative mating: mating between animals that look or perform alike.

Autosomes: the chromosomes of the animal in contrast to the sex chromosomes.

Back-cross: cross between an F_1 (first cross) and either of its parents.

Biometry: the science of statistics related to biology.

Blood lines: general term used to describe relationships.

Blood typing: classifying the blood group of an animal and establishing parentage from this.

Broody: the stage in the fowl's productive life when it has finished laying and sits on eggs.

Breeding Value: assessment of the future genetic potential of an animal.

Broiler: bird bred for meat production.

Castration: removal of the gonads (testicles or ovaries).

Cell: basic unit of all living tissue.

Cervix: opening into the uterus or womb.

Chromatids: the parts of the chromosome after it has split down its length in cell division.

Chromatin: the material in the cell nucleus from which the chromosomes are formed.

Chromosome: the thread-like structure in the cell nucleus that carries the genes.

Cleft palate: divided palate causing difficulty in suckling milk.

Collateral: a related individual from a common ancestor.

Colostrum: the first milk produced by the dam after birth.

Common ancestor: an ancestor that appears on both sire's and dam's side of the pedigree.

Common environment: environment in which all the animals in a group or population are run.

Compensatory growth: growth that occurs after a period of under-feeding when lost weight is regained.

Conditioned reflex: action that triggers or releases another function.

Consanguinuity: of common blood or descent.

Contemporaries: animals born at the same time and similarly reared and treated.

Controls: populations of animals that are unselected and against which other selected populations are compared.

Correlation: statistical term to describe the relationship or association between characters.

Crossing-over: the mutual exchange between homologous pairs of chromosomes.

Crossbred: progeny from crossing two breeds, lines or strains.

Culling: removal of poor animals from a population.

Deviation: a statistical measure of the difference of a particular observation from the mean of the group of observations in which it is found.

DNA (deoxyribonucleic acid): the chemical compound that makes up the basic structure of the gene.

Diploid: the normal double chromosome state which has the correct number of chromosomes for the species.

Distribution: statistical term to describe the variation or spread in a series of observations.

Divergence: separation of factors in different directions.

Dizygous: originating from two separate eggs. Non-identical twins.

Docking: removal of an animal's tail.

Dominant: condition where one allele masks the effect of the other (recessive) allele.

Draft off: remove certain animals from a group.

Dropsical: condition where tissues fill with fluid.

Dynamometer: instrument for measuring energy used in pulling.

Dystocia: difficulty in the process of giving birth.

Embryo: an organism in the early stages of development in the uterus (mammal) or shell (bird).

Embryo transfer (ET): technique of removing an embryo from one female (donor) and implanting it into another female (recipient).

Enzymes: substances produced by the cell that trigger off other reactions.

Epistasis: where one allele appears to be dominant over another allele but they are not at the same locus.

Eye: term used to describe the instinct and action of a sheepdog when working sheep.

Fallopian tube: tube that extends from the uterus to the ovary and in which the ova are fertilised.

Fecundity: a measure of the number of offspring born and reared by the dam.

Fertility: a measure of the ability of the female to conceive and produce offspring, or of the male to fertilise the female.

Foetus: unborn animal inside uterus.

Follicle: structure in the skin from which hair fibres grow.

F_1: the first filial (or daughter) generation or the first cross.

F_2; F_3; *etc:* are subsequent filial generations or the second or third crosses.

Frequency: statistical term to describe the number of times an observation or a gene occurs in a population.

Gamete: the reproductive cells (male sperm and female egg) that unite to produce the offspring or zygote.

Gene interaction: where the same traits are affected by more than one gene.

Generation interval: the average age of the parents when the offspring are born.

Genes: the basic units of inheritance.

Genetic drift: the change in genotypes that occur under the influence of random effects in the environment.

Genetic engineering: the science of modifying the genetic constitution of the animal directly through manipulation of genes.

Genetic isolate: a strain or line that differs genetically from other strains or lines.

Genotype: the genetic make-up of the animal.

Germ cell: a gamete.

Germ plasm: the genetic material in an animal or a population.

Haemophilia: a disease in which blood fails to clot.

Half-bred: an F_1 (first cross) between two parent breeds.

Haploid: half the genetic constitution of the animal as found in the sperm and the egg. Contrasts with the diploid or normal double state of the chromosomes.

Heritability: the strength of inheritance of the trait. Denoted by h^2.

Heterosis: condition that can occur when animals of different genetic constitution are crossed.

Heterozygote: organism that received unlike alleles for a specific locus from its parents.

Hierarchy: structure which shows the genetic composition and organisation of a population.

Histogram: diagram showing variation in a number of observations.

Hogget: sheep approximately six to twelve months of age.

Homologous: of common descent – chromosomes that occur in pairs.

Homozygote: organism that received like alleles for a given locus from its parents.

Hormone: secretion from special glands within the animal that affect various functions.

Inbreeding coefficient: the rate at which heterozygosity is reduced (or homozygosity is increased) per generation in the population.

Inbreeding depression: lowered performance that arises through increased inbreeding.

Index: a computed assessment or estimate of an animal's genetic value based on a number of different traits.

Inversions: where the order of genes on chromosomes can be reversed.

Joining: putting males and females together. *Mating* is used to describe when females are actually served by the males.

Karotyping: examination of chromosomes.

Kemp: very coarse fibre in a fleece, heavily medulated and shed annually.

Killing-out percentage (dressing %): the (carcass weight ÷ liveweight) × 100.

Let-down: release of milk from the udder caused by a hormone (oxytocin).

Lethal gene: gene which when expressed can cause death.

Linkage: association of genes that appear to be inherited together.

Locus: point occupied by a gene on a chromosome. The plural is *loci*.

Maintenance: the feed cost of keeping the animal alive and carrying out its basic functions.

Mean (or average): calculated as the total divided by the number of observations.

Medulla: cavity or hollow inside a fibre.

Meiosis: cell division in the reproductive (germ) cells where the chromosomes are reduced from the diploid (double) state to the haploid (half) state.

Mitosis: cell division of body cells where each new daughter cell receives the normal diploid set of chromosomes.

Mongrel: a crossbred that is unacceptable or unplanned.

Monoploid: the true diploid of the organism.

Monozygous: originating from one egg. Identical twins.

Mutation: a change in the genetic material (germ plasm) of the organism.

Natural selection: selection not influenced by man.

Nicking: term used to describe a successful cross.

Normal distribution: the bell-shaped or Gaussian distribution that describes the variation in the characteristics in many populations.

Nucleus: the central part of the cell in which the genetic material – chromosomes and genes – are found. The plural is *nucleii.*

Objective trait: one that can be defined and measured in a precise way.

Oestrus: the time of heat in the female when she will accept mating by the male.

Ovary: the female organ that produces the ova (eggs).

Over-dominance: where the heterozygote is superior to either homozygote.

Parameter: statistical term for a measure or estimate.

Pathogens: disease organisms.

Pause: rest period in bird's laying cycle.

Peck order: order of social dominance in a group of animals or birds.

Pendulous: udder that hangs down excessively due to weakness of suspensory ligaments.

Performance test: method of evaluating an animal on its own performance.

Perinatal: around birth.

Phenotype: the outward expression of the animal's genetic make-up (genotype).

Placenta: structure in the uterus of the dam through which the foetus is nourished during pregnancy via the umbilical cord. It is expelled at birth. Also called the *afterbirth.*

Plateau: level reached after a period of selection when no further progress is apparent.

Pleiotropy: where one gene has a simultaneous effect on more than one trait.

Point mutation: mutation that takes place within a gene.

Polar body: cell cast off from the ovum during the maturation process of cell division.

Polygenic: concerned with many genes.

Polyploidy: having more than twice the normal number of chromosomes.

Population: group of individual animals.

Pre-potency: the ability of an animal to produce offspring like itself.

Progeny test: evaluation of an animal by examining the performance of its progeny.

Protoplasm: material found in the main body of the cell.

Puberty: sexual maturity.

Quarter-bred: cross obtained when the F_2 is mated back to one of the original parents. Depending on the parent breed of interest it may be called a three-quarter bred.

Random, randomisation: to arrange according to chance and remove bias caused by any other factors.

Random breeding: where each male has an equal chance of mating with each female.

Recessive: an allele that is masked by another (dominant) one.

Reciprocal cross: a cross where the previous parent breeds; strains or individuals have been reversed from male A × female B to female A × male B.

Reduction division: the stage in cell division when the chromosome number is halved.

Reference animal: an animal (usually a sire) against which others can be compared.

Regression: statistical term to describe how one variable changes as another one changes.

Relative Economic Value (REV): an estimate of the relative value (in money terms) of a number of different traits.

Repeatability: statistical term used to describe the chances of traits being repeated.

S/P ratio: ratio between the secondary and primary follicles in the skin.

Scatter diagram: shows the scatter or distribution of a series of paired observations for a trait in a group of animals.

Scour: the process of washing wool to remove grease and contaminants.

Screen: to select from a large population those animals that have approved specifications to form an elite or nucleus group.

Scurs: small horns that are not part of the animal's skull but are attached to the skin.

Segregation: separation of the alleles of a pair in the formation of the germ cells so that each allele goes to separate gametes.

Selection differential: the difference between the mean of the selected parents and the mean of the population from which they came.

Semen: the male sperm and fluids produced in the testicles and other glands of the male's reproductive system.

Settling-in period: a period at the start of a performance test when the animals adjust to the feeding and management system.

Sex chromosomes: chromosomes that are concerned specifically with the inheritance of sex.

Sex limited: traits that are limited by the sex of the animal i.e. they can only be expressed in one sex.

Sex linked: traits that are carried on the sex chromosomes.

Sigmoid curve: an S-shaped curve used to describe growth from conception to maturity.

Sibling (sib): offspring of the same parents, but not necessarily born all at the same time. Full sibs have both parents in common while half-sibs have only one parent in common.

Skewed: a distribution that is distorted and has only one tail.

Spermatozoa (sperm): the male gametes.

Standard deviation: statistical term that describes the variation in a trait around a mean value.

Stud: a term used to describe a breeder or his flock or herd that is registered (pedigree) with an official breed association.

Subjective trait: one that cannot be defined or expressed in a precise way.

Suint: exudate from the sweat gland in the skin.

Superovulation: act of stimulating the ovary to produce more eggs than normal.

Teaser: vasectomised male that acts as a normal male but cannot pass viable sperm.

Telegony: the unfounded belief that a dam's previous offspring can affect those born later and by a different sire.

Tetraploid: having twice the normal number of chromosomes.

Threshold: base above which expression of the gene will be seen and below which it will not.

Top-cross: cross by a sire from a new blood-line of the same breed.

Trait: character or characteristic.

Translocation: transfer of a part of a chromosome to another part of the same chromosome.

Triploid: having three times the haploid number of chromosomes for the species.

Truncation: a point in a distribution after which animals are selected or culled.

Two-tooth: a sheep showing two permanent incisors. Approximately eighteen months of age.

Ultrasonics: equipment that uses high frequency sound waves to assess what is below the skin of a live animal such as fat layers and muscle area.

Uterus: the female organ in which the foetus develops during pregnancy.

Variation: measure of animal differences in a group or population.

Yearling: animal of approximately one year of age.

Zygote: product of the union of two gametes.

Further Information

Textbooks are a major source of information but they have the disadvantage that the information in them rapidly becomes dated unless they are regularly reprinted. This is especially the case with results of current research which are reported in research papers in scientific journals.

Research papers

Research workers publish their findings in the recognised scientific journals. Examples of journals that would carry the latest animal breeding research information are the following:

Africa: *African Journal of Agricultural Sciences.* (Afr. J. Agric. Sci.)

Australia: *Australian Journal of Experimental Agriculture and Animal Husbandry.* (Aust. J. Agric. Anim. Husb.)

Britain: *Animal Production.* (Anim. Prod.)

Canada: *Canadian Journal of Animal Science.* (Canad. J. Anim. Sci.)

Europe: *Livestock Production Science.* (Livest. Prod. Sci.)

India: *Indian Journal of Animal Production.* (Indian J. Anim. Prod.)

Japan: *Japanese Journal of Breeding.* (Jap. J. Breed.)

New Zealand: *New Zealand Journal of Agricultural Research.* (N.Z. J. Agric. Res.)

Pakistan: *Pakistan Journal of Animal Sciences.* (Pakistan J. Anim. Sci.)

Philippines: *Philippine Journal of Animal Industry.* (Philipp. J. Anim. Ind.)

Scandinavia: *Acta Agriculturae Scandinavica.* (Acta Agric. Scand.)

South Africa: *South African Journal of Animal Science.* (South Afr. J. Anim. Sci.)

USA: *Journal of Animal Science.* (J. Anim. Sci.)

Zimbabwe: *Zimbabwe Journal of Agricultural Research.* (Zimbabwe J. Agric. Res.)

General: *World Review of Animal Production.* (World Rev. Anim. Prod.)

The abbreviation for the journal is shown in brackets.

If readers want a paper from a journal, they can obtain a reprint or offprint by writing to the author. The address of the senior author of the paper (which appears first in a joint-author paper) is given in the title and there is usually no charge for reprints.

Readers can also obtain information on the papers being published in scientific journals through a journal called *Current Contents.* Here the contents pages of the latest available issues of journals are listed. The address of the senior author of each paper is given in an index. Animal breeding is covered by the Agriculture, Biology and Environmental Sciences section. Further information about *Current Contents* can be obtained from The Institute for Scientific Information, 3501 Market Street, University City Science Centre, Philadelphia, 19104, USA.

Abstracting journals

To help readers who may not be able to obtain the many scientific journals available around the world, and especially those published in foreign languages, the Commonwealth Bureau of Animal Breeding and Genetics publishes an abstracting journal called *Animal Breeding Abstracts (ABA).*

Staff at the Bureau's headquarters in Edinburgh scan papers from over one thousand world scientific journals from which they prepare abstracts in English. These are short statements giving the main findings of the work. Anyone can subscribe to *ABA* and details can be obtained from the Director (Dr J. D. Turton, Commonwealth Bureau of Animal Breeding and Genetics, King's Buildings, West Mains Road, Edinburgh EH9 3JX, Scotland).

Review articles

Animal Breeding Abstracts regularly carries review articles written by authorities in the field. Here the author gathers all the research together, comments on its relevance, draws conclusions and lists all the references cited.

A review article is an ideal place to start gaining further information, as a good reviewer will draw conclusions for you. Here are some examples or review articles that have appeared in *ABA* in the 1960–84 period.

Information on obtaining copies of these review articles can be obtained from the Director of the Bureau.

Al-Murrani, W. K. (1974) 'The limits to artificial selection'. *ABA* **42**, 587.

Barlow, R. (1978) 'Biological ramifications of selection for pre-weaning growth in cattle'. A review. *ABA* **46**, 469.

Barton, R. A. (1967) 'The relation between live animal conformation and the carcass of cattle'. *ABA* **35**, 1.

Bird, P. J. W. M. and Mitchell, G. (1980) 'The choice of discount rate in animal breeding investment appraisal'. *ABA* **48**, 499.

Bowden, Valerie. (1982) 'Type classification in dairy cattle: a review.' ABA **50**, 147.

Bowman, J. C. (1966) 'Meat from sheep'. *ABA* **34**, 293.

Dickerson, G. E. (1962) 'Random sample performance testing of poultry in the USA'. *ABA* **30**, 1.

Dýrmundsson, Ó. R. (1973) 'Puberty and early reproductive performance in sheep'. I. Ewe lambs. *ABA* **41**, 273.

Ibid (1973)
II. Ram lambs. *ABA* **41**, 419.

Edey, T. N. (1969) 'Prenatal mortality in sheep – a review'. *ABA* **37**, 173.

Fredeen, H. T. (1965) 'Genetic aspects of disease resistance'. *ABA* **33**, 17.

Fredeen, H. T. (1967) 'Where should we be going in animal breeding research?'. *ABA* **35**, 23.

Gordon, I. (1976). 'Controlled breeding in cattle. II. Pregnancy testing, control of calving, reduction of the calving interval, induction of twinning, breeding at younger age, future developments'. *ABA* **44**, 451.

Hendy, C. R. C. and Bowman, J. C. (1970) 'Twinning in cattle'. *ABA* **38**, 22.

Hinks, C. J. M. (1978) 'The use of centralised breeding schemes in dairy cattle improvement'. *ABA* **46**, 291.

Hughes, J. G. (1976) 'Short term variation in animal liveweight and reduction of its effect on weighing'. *ABA* **44**, 111.

Johansson, I. (1964) 'The relation between body size, conformation and milk yield in dairy cattle'. *ABA* **32**, 421.

Joubert, D. M. (1963) 'Puberty in female farm animals'. *ABA* **31**, 295.

King, J. W. B. (1970) 'Organisation and practice of pig improvement in European countries'. *ABA* **38**, 523.

Land, R. B. (1974) 'Physiological studies and genetic selection for sheep fertility'. *ABA* **42**, 155.

López-Fanjul (1974) 'Selection from crossbred populations'. *ABA* **42**, 403.

Mason, I. L. (1966) 'Hybrid vigour in beef cattle'. *ABA* **34**, 453.

Mason, I. L. (1971) 'Comparative beef performance of the large cattle breeds of western Europe'. *ABA* **39**, 1.

Miller, R. H. and Pearson, R. E. (1979) 'Economic aspects of selection'. *ABA* **47**, 281.

Morris, C. A. and Wilton, J. W. (1977) 'The influence of body size on the economic efficiency of cows – a review'. *ABA* **45**, 139.

Morris, C. A. (1980) 'A review of relationships between aspects of reproduction in beef heifers and their lifetime production. I. Associations with fertility in the first joining season and with age at first joining'. *ABA* **48**, 655.

Moule, J. R. (1970) 'Australian research into reproduction in the ram'. *ABA* **38**, 185.

Newton-Turner, H. (1969) 'Genetic improvement of reproduction rate in sheep'. *ABA* **37**, 545.

Newton-Turner, H. (1977) 'Australian sheep breeding research'. *ABA* **45**, 9.

Nitter, J. (1978) 'Breed utilization for meat production in sheep'. *ABA* **46**, 131.

Pearson, Lucia and McDowell, R. E. (1968) 'Crossbreeding of dairy cattle in temperate zones – a review of recent studies'. *ABA* **36**, 1.

Pearson de Vaccaro, Lucia (1973) 'Some aspects of the performance of purebred and crossbred dairy cattle in the tropics'. *ABA* **41**, 571.

Rollinson, D. H. L. (1971) 'Further development of artificial insemination in tropical areas'. *ABA* **39**, 623.

Ryder, M. L. (1980) 'Fleece colour in sheep and its inheritance'. *ABA* **48**, 305.

Sheridan, A. K. (1981) 'Crossbreeding and heterosis.' ABA **49**, 131.

Turton, J. D. (1981) 'Crossbreeding of dairy cattle – a selective review' ABA **49**, 293.

Yüksel, E. (1979) 'Genetic aspects of the efficiency of food utilization in some farm and laboratory animals'. *ABA* **47**, 499.

Annotated bibliographies

The Commonwealth Bureau of Animal Breeding and Genetics also prepares bibliographies for users. This is where someone requests a copy of all the published work on a specific subject and receives a copy of the abstract of each paper.

Duplicate copies of these bibliographies are available for sale from the Director of the Bureau who will also provide a complete list of those available. These bibliographies are regularly updated.

Examples

Number	Subject	Period covered	No. of references
277A	Selection in dairy cattle	1973–8	93
341	AI and reproduction in the buffalo	1970–76	231

Information retrieval systems

Information can now be stored electronically in data bases. Here large computers have all the material on file and it can be accessed by users. Currently, the method is to consult a librarian with the expertise and equipment to access the appropriate data base holding animal breeding information.

The user's main job is to define precisely what information is wanted so that it can be extracted. Unless you can define your needs very carefully, you will obtain a mass of information that you do not need and this will waste time and add confusion.

The two data bases most appropriate for animal breeding information are CAB ABSTRACTS and AGRIS. The CAB ABSTRACTS data base is available through the Lockheed Research Laboratory in Palo Alto, California, USA. Contact the Director of the Commonwealth Bureau of Animal Breeding and Genetics for details. AGRIS is based in Vienna and is owned by the Food and Agriculture Organisation (FAO).

Users throughout the world can access these information sources. Your nearest librarian should be consulted for information on these and other retrieval systems.

General breeding advice

In most countries, the government's agricultural advisory service is the best place to go for advice. These organisations will be able to provide further information on where help can be obtained. They may have specialist breeding advisers.

Index